TECHNICAL MANUAL
NO. 9-1005-224-10

***TM 9-1005-224-10**
T.O. 11W2-6-4-11
HEADQUARTERS
DEPARTMENTS OF THE ARMY
AND AIR FORCE
Washington D.C., 2 April 1998

Operator's Manual for MACHINE GUN, 7.62-MM, M60

REPORTING ERRORS AND RECOMMENDING IMPROVEMENTS
You can help improve this manual. If you find any mistakes or if you know of a way to Improve the procedures, please let us know. Mail your letter or DA Form 2026 (Recommended Changes to Equipment Publications and Blank Forms) direct to: Director, Armament and Chemical Acquisition and Logistics Activity, ATTN: AMSTA-AC-NML, Rock Island, IL 61299-7630. A reply will be furnished to you.

*This manual supersedes TM 9-1005-224-10, dated 30 July 1985; TM 9-1005-224-10HR, dated 18 May 1979; TB 9-1005-224-50-1, including all changes.

WARNING SUMMARY

Be sure to clear weapon before starting inspection.

If cover is opened on a hot cartridge (hot barrel), an open cover cookoff could occur and result in serious injury or death. Evacuate area for 15 minutes and then do REMEDIAL ACTION. Refer to WP 00011 00.

Never reload a runaway machine gun until it has been repaired. Be sure machine gun is cleared before moving it.

When removing a stuck unfired cartridge, stay clear of the muzzle. Do not allow cartridge to contact any hard surface. Cartridge may fire on contact. Remove a stuck unfired cartridge using the same procedures for removing a stuck or ruptured cartridge case.

Before field stripping check that the bolt is in forward position.

To prevent possible bodily injury, or aircraft/weapon damage, personnel should not stand the weapon on its barrel assembly when disassembling or assembling the weapon.

Bolt assembly in under spring tension; it can twist and injure your hand.

Barrels issued for a specific gun will not be changed from gun to gun. Each barrel and bolt assembly should be tagged during cleaning with the serial number of the receiver (gun) and should remain together as initially assigned. Interchanging barrels/bolts may result in injury or death.

The climate temperature in different regions will make a difference as to what constitutes a hot gun. A hot, sunny day can cause a cookoff within 50 rounds, weapon and ammunition in the sun.

Insert the hinge pin from one side and the hinge pin latch from the opposite side.

After unloading dummy round, be sure the barrel is clear.

Ammunition to be used in machine gun is shown in WP 0031 00.

Keep weapon pointed at target - down range/impact area.

Keep safety on 'S' (safe) until your ready to fire.

Always check the chamber/bore after clearing the weapon

Check barrel bore and chamber before firing.

Do not interchange bolt/barrel assemblies from one weapon to another. This may cause injury or death to personnel.

Change hot barrels with your heat protective mitten only or by holding bipod legs. (Refer to WP 0017 00 for additional hot barrel procedures.)

Do not interchange flat leaf springs from one model weapon to another.

Check that assigned and spare barrels have been headspaced and tagged to your receiver. Rotate usage of the barrel.

Do not fire blank ammunition toward personnel within 20 feet of the muzzle. Fragments of a closure wad or particles of unburned propellant might inflict injury within that range.

Dry cleaning solvent is FLAMMABLE and TOXIC and must be kept away from open flames and used in a well ventilated area. Use of rubber gloves is necessary to protect the skin when washing machine gun parts.

Appropriate eye protection is recommended when cleaning your weapon and/or its parts.

For first aid information see FM 21-11.

TABLE OF CONTENTS

CHAPTER 1

INTRODUCTION

GENERAL INFORMATION 0001 00

SCOPE

Type of Manual: Operator's Manual.

Model Numbers and Equipment Names: M60, 7.62mm, Machine Gun, and M60D, 7.62mm, Machine Gun.

Purpose of Equipment:

The M60, 7.62mm, Machine Gun is a general purpose weapon capable of being fired from several mounts or handheld. The weapons primary use is for ground operations.

The M60D, 7.62mm, Machine Gun is a general purpose weapon capable of being fired from several mounts. The weapons primary use is for support of ground operations. The M60D is an aircraft door-mounted or vehicle-mounted machine gun.

MAINTENANCE FORMS AND RECORDS

Department of the Army forms and procedures used for equipment maintenance will be those prescribed by DA PAM 738-750, the Army Maintenance Management System (TAMMS).

REPORTING EQUIPMENT IMPROVEMENT RECOMMENDATIONS (EIR's).

If your machine needs improvement, let us know. Send us an EIR. You, the user, are the only one who can tell us what you don't like about your equipment. Let us know why you don't like the design or performance. Put it on an SF 368 (Quality Deficiency Report). Mail it to us at Commander, US Army Armament Research, Development and Engineering Center, ATTN: AMSTA-AR-QAW-A, Rock Island, IL 61299-7630. We'll send you a reply.

EQUIPMENT CHARACTERISTICS, CAPABILITIES AND FEATURES.

The M60 and M60D are air-cooled and have fixed headspace allowing quick barrel changes for cooling and maintenance when required. In order to extend the life of the barrels, retain accuracy, and allow for continuous firing over long periods of time, two barrel assemblies are issued with each gun. The machine guns are gas-operated and fire from open bolt position.

LOCATION AND DESCRIPTION OF MAJOR COMPONENTS.

M60 and M60D Machine Gun.

A **COVER ASSEMBLY** - Positions and holds cartridge in place for stripping, feeding link belt, and chambering rounds.

B **COCKING HANDLE ASSEMBLY** - Provides a means to manually move the bolt assembly to the rear.

C **TRIGGER MECHANISM AND GRIP ASSEMBLY (M60 ONLY)** - Controls the firing of the machine gun.

D **SEAR AND SAFETY HOUSING (M60D ONLY)** - Controls the firing of the machine gun.

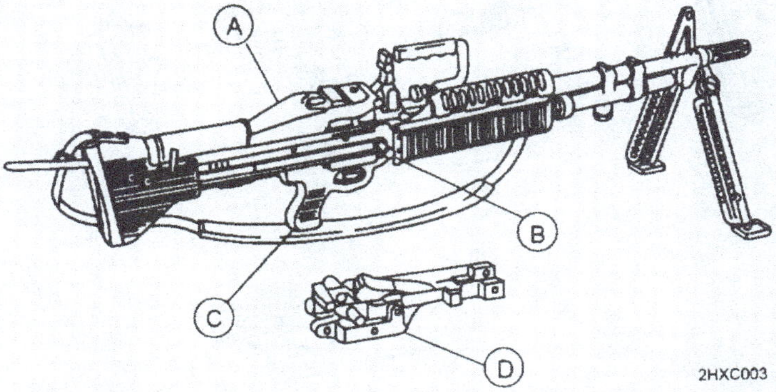

LOCATION AND DESCRIPTION OF MAJOR COMPONENTS (cont)

E **BARREL AND BIPOD ASSEMBLY** - The barrel houses cartridges for firing. The bipod assembly provides a semistable platform when the machine gun is fired from the prone position.

F **CARRYING HANDLE ASSEMBLY** - Provides a means to carry the machine gun with one hand. The carrying handle assembly folds down when the rear sight is used and the machine gun is fired.

G **RECEIVER ASSEMBLY** - Supports all major components. Houses internal parts and, through a series of cam ways, controls operation of weapon.

H **SHOULDER GUN STOCK (M60 ONLY)** - Provides a suitable surface to stabilize the weapon against the shoulder while firing the machine gun from any position except from the hip.

I **GRIP AND TRIGGER ASSEMBLY (M60D ONLY)** - Provides handles to move machine gun and houses the machine gun trigger.

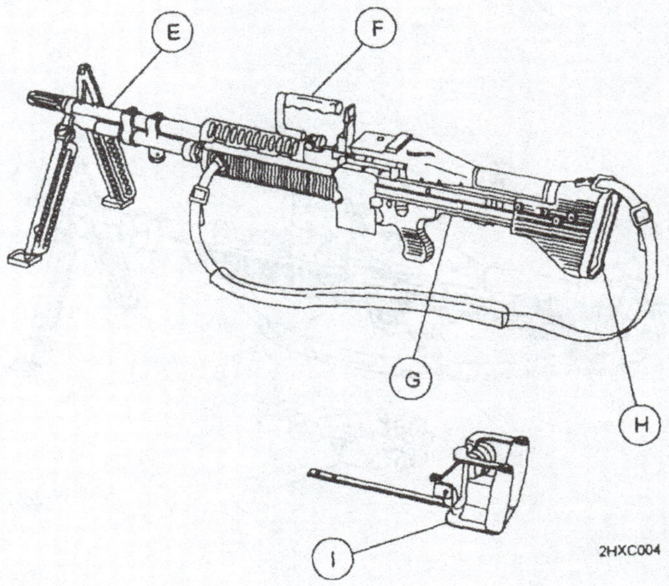

LOCATION AND DESCRIPTION OF MAJOR COMPONENTS (cont)

J **CARTRIDGE FEED TRAY AND HANGER ASSEMBLY (M60 ONLY)** - Guides cartridges for positioning and feeding. The hanger assembly supports the bandoleer.

K **CARTRIDGE TRAY ASSEMBLY (M60D ONLY)** - Guides cartridge for positioning and feeding.

L **REAR SIGHT ASSEMBLY (M60 ONLY)** - Provides a means to aim the machine gun at the target with accuracy. The sight adjusts horizontally as well as vertically.

M **REAR SIGHT ASSEMBLY (M60D ONLY)** - Provides a means to aim the machine gun in the general area of the target. The rear sight is nonadjustable.

N **FOREARM ASSEMBLY (M60 ONLY)** - Provides a hand hold when firing from the hip or from a standing or kneeling position.

O **SMALL ARMS SLING (M60 ONLY)** - Used for support during assault firing and for transport.

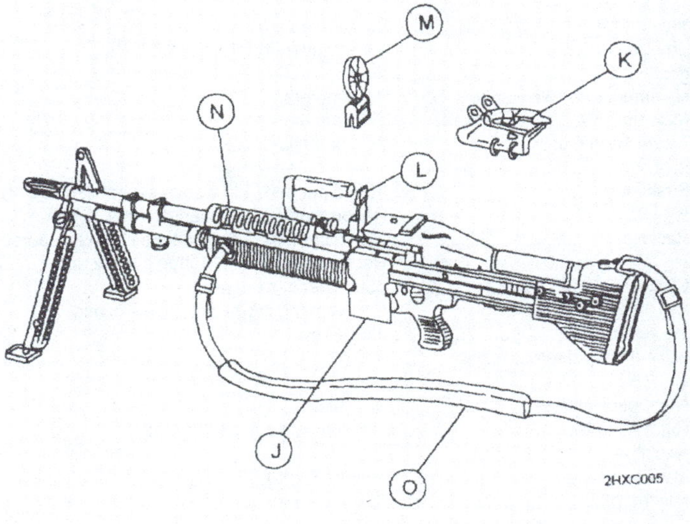

DIFFERENCES BETWEEN MODELS - Differences between the M60 and M60D machine gun are listed below.

Assembly/Component	M60	M60D
Tray and hanger assembly	X	
Cartridge feed tray assembly		X
Dust and moisture seal boot		X
Forearm assembly	X	
Grip and trigger assembly		X
Gun adapter		X
Magazine hanger assembly		X
Rear sight (adjustable)	X	
Rear sight (nonadjustable)		X
Receiver assembly	X	
Receiver assembly		X
Sear and safety housing assembly		X
Sear assembly link and spring		X
Shoulder gun stock	X	
Small arms sling	X	
Trigger mechanism grip assembly	X	
Quick release pin		X

EQUIPMENT DATA

M60 Machine Gun.

Weight... 23 lb (10.43 kg)
Length..................................... 43.5 in. overall (1.1m overall)
Range:
- Maximum Effective Range 1100meters
- Maximum Range .... 3725 meters
- Tracer Burn Out .... 900 meters or more

Rates of fire:
- Sustained .... 100 rd per min (4 to 6 seconds between burst) Recommend barrel change every 10 min
- Rapid .... 200 rd per min (2 to 3 seconds between burst) Recommend barrel change every 2 min
- Cyclic .... 550 rd per min (approx) Recommend barrel change every min

Muzzle velocity .... 2800 FPS
Capacity of bandoleer .... 100 rds
Rifling:
- Number of lands......................... 4
- Right hand twist........................... One turn in 12 in (30.54 cm)

Trigger pull:
- Maximum.................................. 11.5 lb (5.2 Kg)
- Minimum.......................................6.0 lb (2.7 kg)

EQUIPMENT DATA (cont)

M60D Machine Gun

Weight	25 lb (11.33 kg)
Length	43.5 in overall (1.1m overall)
Range:	
Maximum Effective Range	1100 meters
Maximum Range	3725 meters
Tracer Burn Out	900 meters or more
Rates of fire:	
Sustained	100 rd per min (4 to 6 seconds between burst) Recommend barrel change every 10 min
Rapid	200 rd per min (2 to 3 seconds between burst) Recommend barrel change every 2 min
Cyclic	550 rd per min (approx) Recommend barrel change every min
Rifling:	
Number of lands	4
Right hand twist	One turn in 12 in (30.54 cm)
Trigger pull at sear activator:	
Maximum	20 lb (9.06 kg)
Minimum	10.5 lb (4.75 kg)

CHAPTER 2

OPERATING INSTRUCTIONS

This page intentionally left blank.

DESCRIPTION AND USE OF OPERATOR'S CONTROLS AND INDICATORS 0003 00

M60 MACHINE GUN

COCKING HANDLE. Pull back to move bolt rearward.

COVER LATCH. Holds cover closed. Pull back on lower end of latch to open.

BARREL LOCK LEVER. Locks barrel in place. To unlock barrel, push in on barrel lock lever from left side of weapon and raise lever.

BIPOD LEG PLUNGER. Depress to extend or retract bipod foot.

TRIGGER. Pull rearward to fire weapon. Release to stop firing.

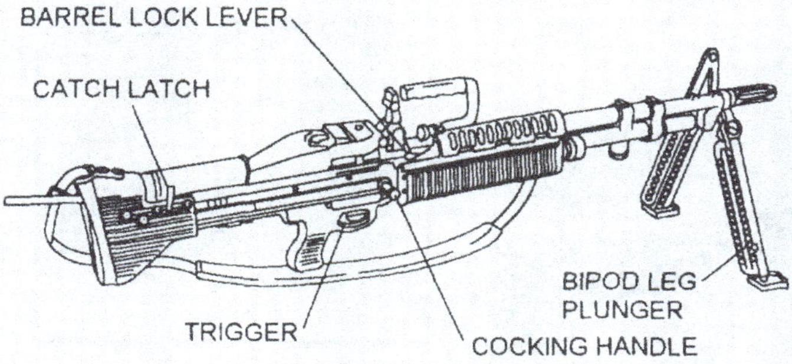

M60

M60 MACHINE GUN (cont)

SAFETY. Move safety up to 'F' (fire) and down to 'S' (safe).

REAR SIGHT. Adjustable to 1100 meters.

- Range Scale - graduated from 300 to 1100 meters.
- Scale Retaining/Adjusting Screw - locks range scale in place.
- Elevating Knob - use for minor (fine) adjustments in range.
- Windage Scale - adjust for left or right wind.
- Rear Sight Aperture - lines up with front sight for long range targets.

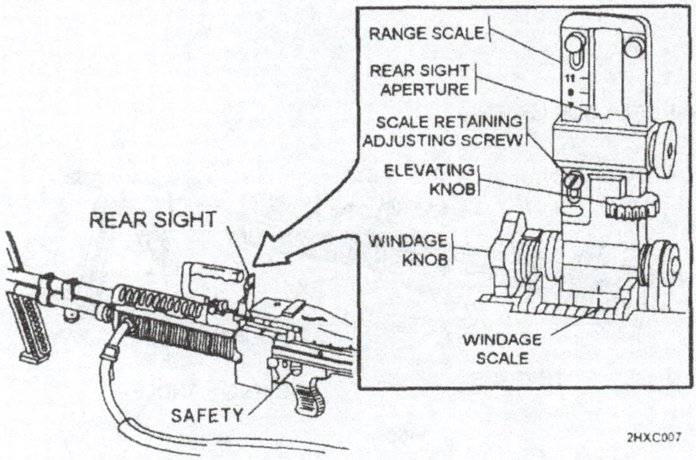

M60D MACHINE GUN

GRIP AND TRIGGER ASSEMBLY. Pull back on both triggers to fire. Release to stop firing.

SAFETY. Push to right for safe 'S', to left for fire 'F'.

REAR AND FRONT SIGHTS. Aim through the ring sight (rear sight) and front sight to line up target. Fire. Adjust range by watching tracers.

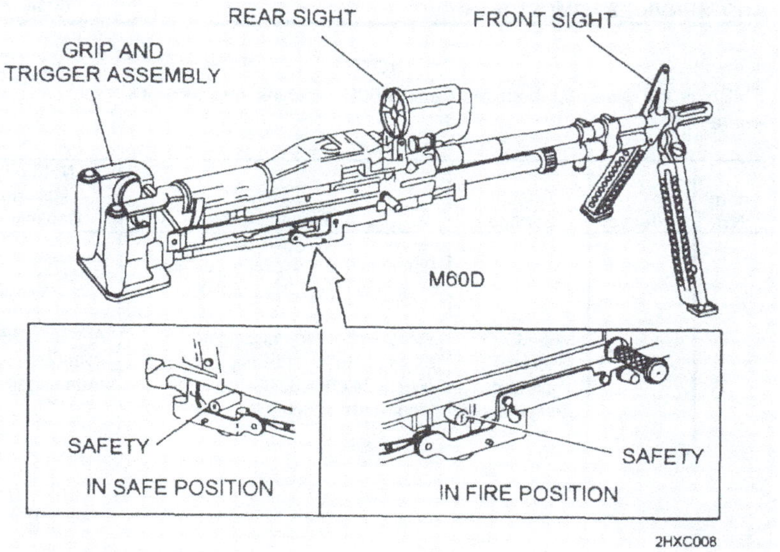

PREVENTIVE MAINTENANCE CHECKS AND SERVICES (PMCS) 0004 00

GENERAL

Always keep in mind the CAUTIONS and WARNINGS when performing your before (B) or during (D) PMCS. The numbers in the item number column shall be used for the "TM Number" column on DA Form 2404, Equipment Inspection and Maintenance Worksheet, in recording results of PMCS. After you operate, perform your after (A) PMCS.

Perform BEFORE PMCS if: (1) you are the assigned operator and the machine gun has been stored and not used for a period of 90 days or(2) you have been issued the machine gun for the first time.

WARNING

Do not interchange barrel assembly or bolt assembly from one machine gun to another. Doing so may result in injury or death.

The climate temperature in different regions will make a difference as to what constitutes a hot gun. A hot, sunny day can cause a cookoff within 50 rounds, weapon and ammunition in sun.

NOTE

PMCS is the same for both M60 and M60D machine gun except where noted.

Item No.	Interval	Item to Check/ Service	Procedure	Not Fully Mission Capable if:
			WARNING Before starting inspection, be sure to clear weapon (WP 0009 00).	
1	B	M60/M60D Machine Gun Tools and Equipment	Visually check for missing or damaged tools and equipment (WP 0034 00/0035 00) including tripod and components.	Any BII/COEI items are missing ordamaged.

Item No.	Interval	Item to Check/ Service	Procedure	Not Fully Mission Capable If:
2	B	Barrel Assemblies	Open cover, pull bolt to rear, push cocking handle forward and lock. Close cover. place on safe (S). Tilt weapon end to end to make sure gas piston slides freely (a clicking sound will be heard as the piston slides In the cylinder). Run a dry swab through the bore to remove excess lubricant or obstructions.	Gas piston will notslide. Obstruction in barrel. Gas cylinder vent, cylinder plug, and cylinder nut loose.
3	B	Trigger Mechanism Grip Assembly (M60)/ Sear and Safety Housing Assembly (M60D)	Make sure flat leaf spring is in place (WP 0018 00). Check that safety is in 'F' position. Pull bolt to rear, move safety to 'S'. Push cocking handle forward and lock and pull the trigger.	Weapon functions with safety at 'S'.

Item No.	Interval	Item to Check/ Service	Procedure	Not Fully Mission Capable If:
4	B	Cover Assembly	Lightly lubricate all moving parts with lubricant on a clean swab (item 6, WP 0038 00).	
5	B	Gun Receiver Assembly	Check for free movement of cocking handle. Check barrel lock lever. Make sure it locks in place.	Cocking handle binds or barrel is loose.
6	D	M60/M60D Machine Gun	Erratic or sluggish firing may indicate carbon buildup. Switch barrels (WP 0018 00). If firing is still erratic or sluggish, field-strip the weapon and perform the necessary maintenance to complete the tire mission.	Weapon ceases to operate.

Item No.	Interval	Item to Check/ Service	Procedure	Not Fully Mission Capable If:
7	D	Small Arms Sling (M60)	Inspect sling for damage and for missing or loose sling swivels.	Swivels loose or missing. Sling is damaged.
8	A	M60/M60D Machine Gun	Fieldstrip, clean, inspect and lubricate entire weapon (WP 0018 00).	
9	A	M60/M60D Machine Gun (Inspection)	Inspect operating rod assembly and barrel assembly for burrs, cracks and chips. Make sure the gas piston slides back and forth freely. Inspect components of trigger assembly for wear. Place safety in 'F' position, sear should move slightly. Inspect breechbolt assembly for defects (e.g., cracks or chips). Reassemble weapon and check for proper functioning.	Weapon fails to function or has damaged or missing parts.

NOTE

The following procedures illustrate the M60 machine gun except where the M60 parts differ from the M60D park, then both weapons are illustrated.

1 Point weapon down range.

2 Safety should be on F.
- M60 - Move safety to F.
- M60D - Push in on F.

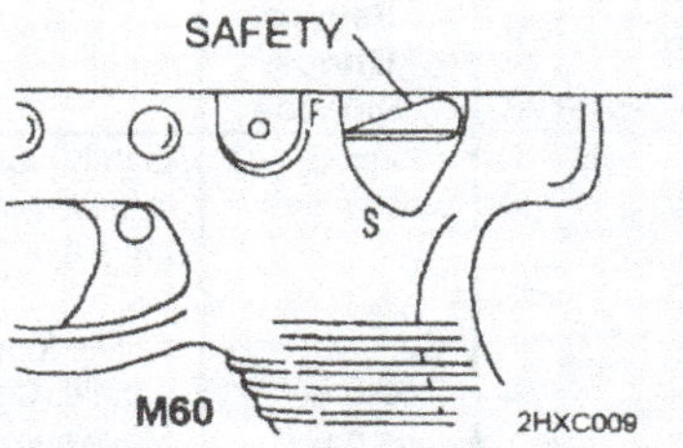

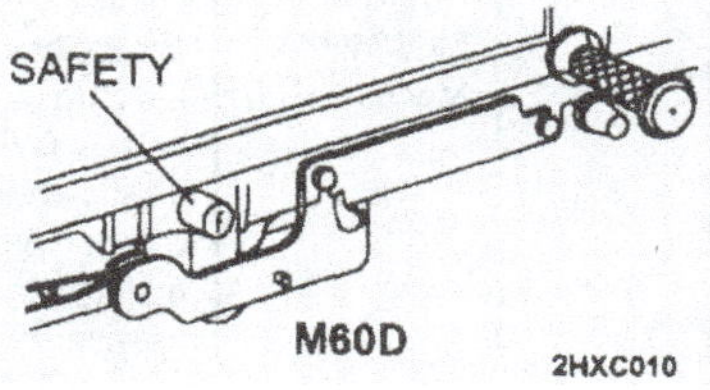

3 Pull cocking handle rearward, then push it forward until it locks.

4 M60 - Move safety to S.
M60 D - Press in on S.

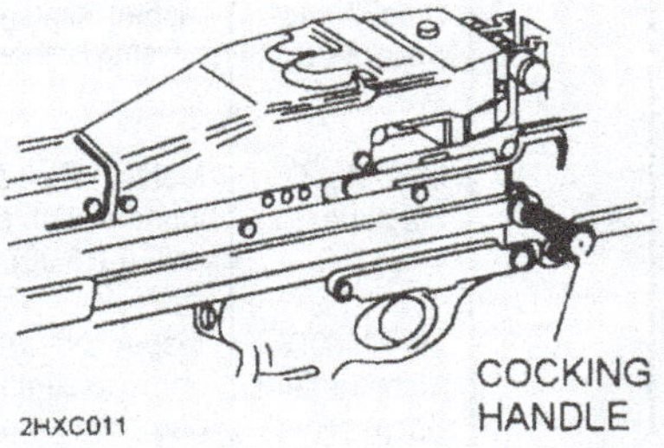

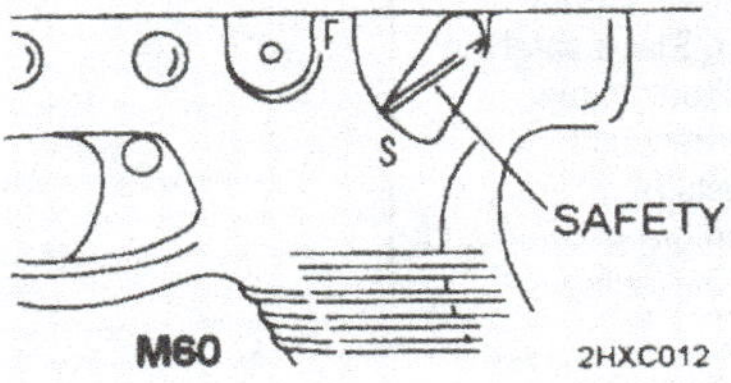

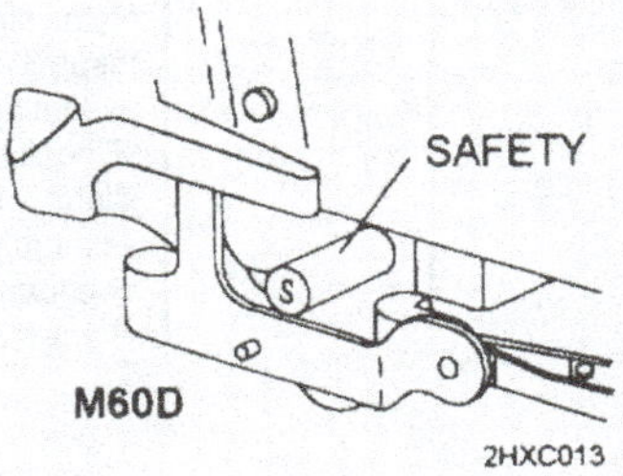

CAUTION

Bolt assembly must be rearward when opening or closing the cover.

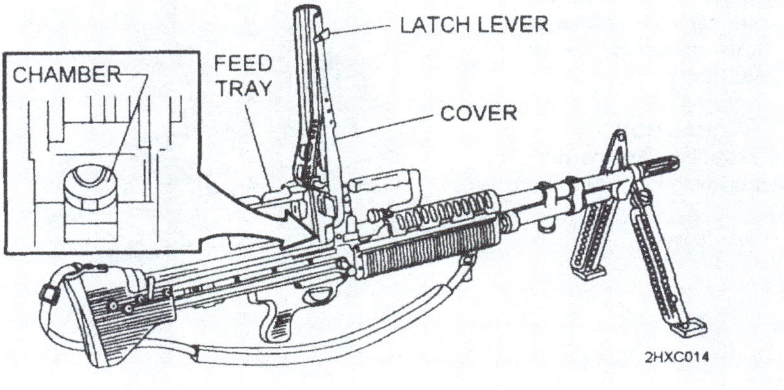

5 Turn latch lever and open cover.

6 Raise cartridge feed tray and check chamber for ammunition.

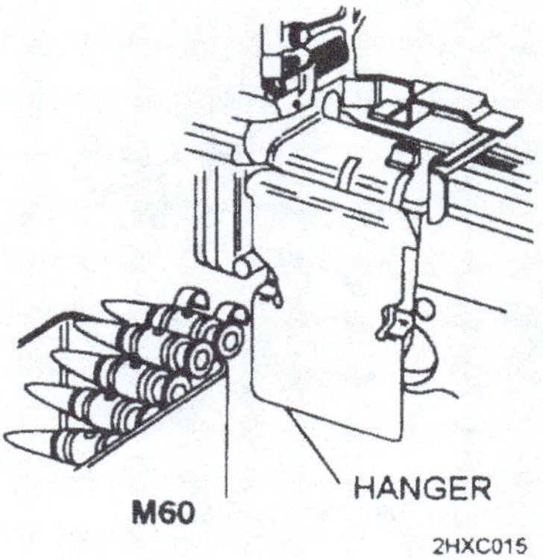

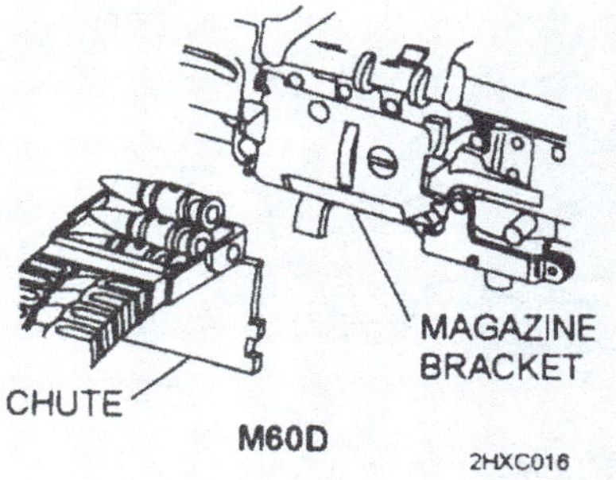

7 M60 - Place bandoleer on bandoleer hanger.

M60D - Attach ammunition chute to magazine bracket.

8 Place ammunition on feed tray, with open side of links down and with first round in the feed groove. (Make sure cartridge retainer pawl holds ammunition link on feed tray.)

CAUTION

Bolt assembly must be rearward when opening or closing the cover.

9 Close cover.

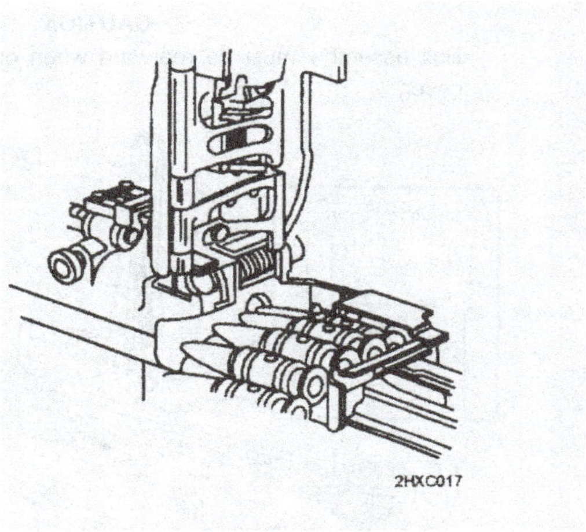

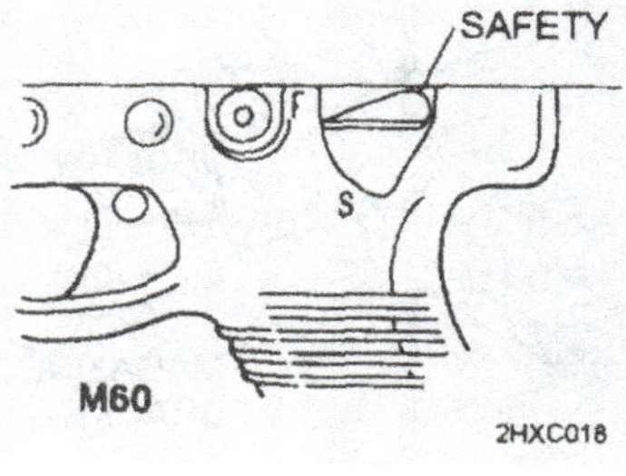

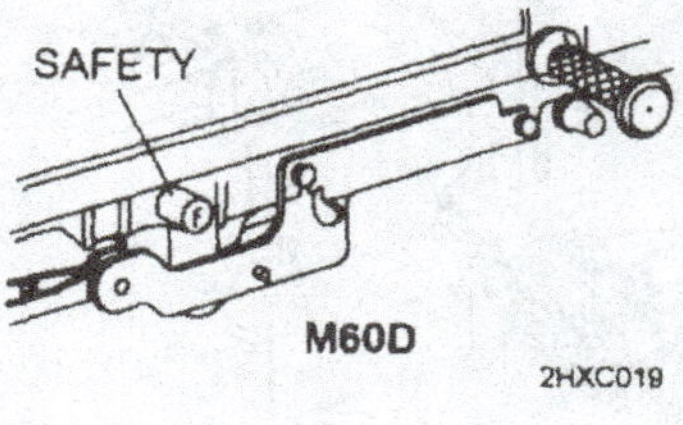

10 M60 - Move safety to 'F'.
M60D - Press in on 'F'.

11 The gun is ready to fire.

1 Unfold front leg and spread rear legs until leg lock engages.

2 Insert pintle assembly. Rotate pintle lock release cam to lock (see FM 23-67).

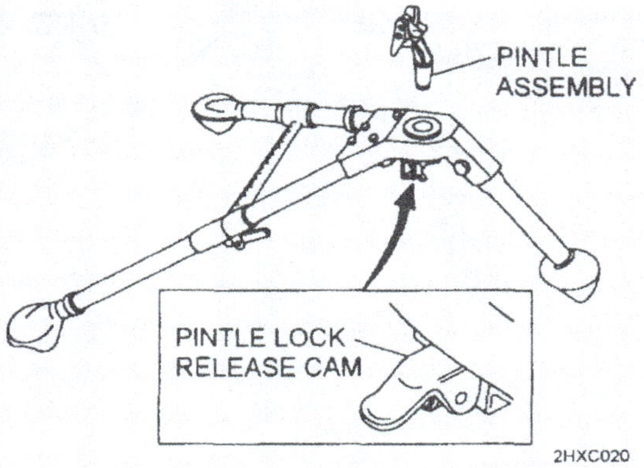

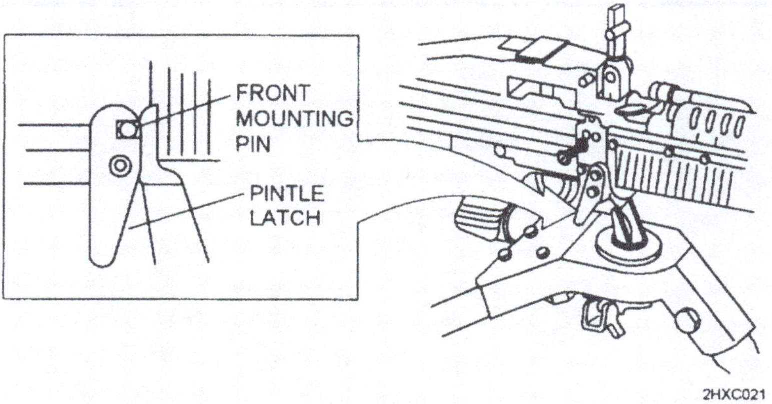

3 Place ends of machine gun's front mounting pin on top of pintle assembly.

4 Press bottom of latch to open pintle assembly. The ends of machine gun mounting pin should lock in place on pintle assembly.

5 Attach gun mount adapter to Traversing and Elevating Mechanism using straight pin and clips.

6 Center (zero) traversing hand-wheel and elevating handwheel on traversing and elevating mechanism assembly (see FM 23-67).

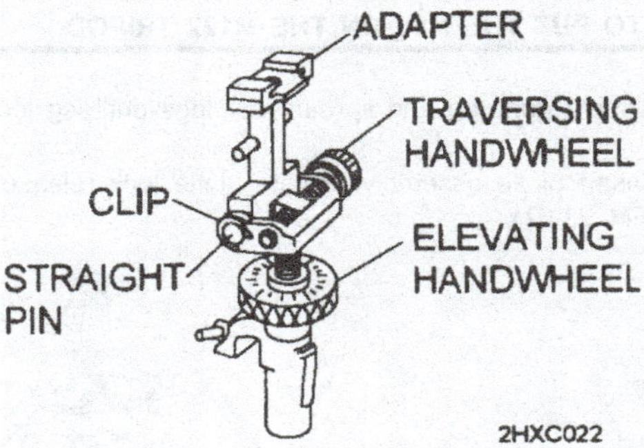

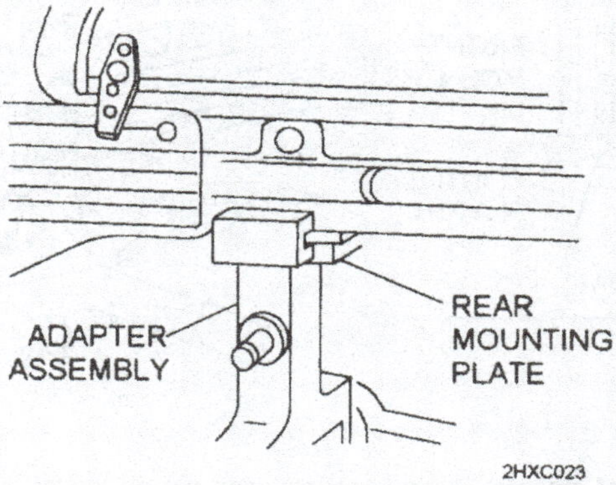

7 Slide adapter assembly on machine gun's rear mounting plate from the rear to the front.

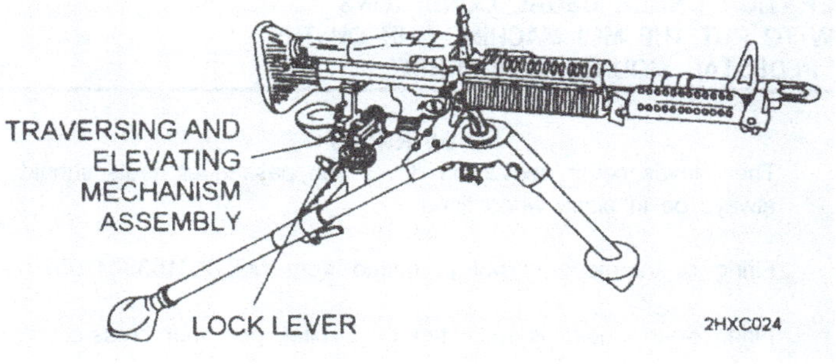

8 Position traversing and elevating mechanism assembly on traversing bar. Rotate lock lever to secure.

OPERATION UNDER USUAL CONDITIONS - HOW TO PUT THE M60 MACHINE GUN ON THE M6 PEDESTAL MOUNT/ M197 MOUNT 0007 00

NOTE

For mounting instructions see TM 9-1005-245-13&P.

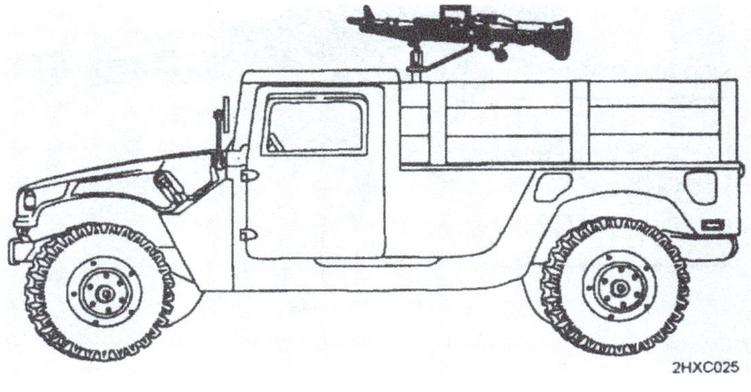

WARNING

The canvas cover above the driver and passenger seats should always be in place when firing.

Firing on-the-move is not permitted from M998/M1038/M1097.

Firing on-the-move is restricted to 5 miles per hour cross-country and 10 miles per hour on improved roads when mounted on M1025 and M1026 HMMWVs (Armament Carriers).

Ground personnel should not be within 10 meters of the vehicle when firing.

CAUTION

When the weapon is not in use/manned it should be secured in the travel lock. When not manned for extended cross-country travel weapon should be removed from the mount,

1 Ensure bipod legs are in the rearward locked position and the M197 pintle latch is rearward.

2 Press bottom of latch to open the pintle assembly. The ends of the machine gun mounting pin should lock in place on the pintle assembly.

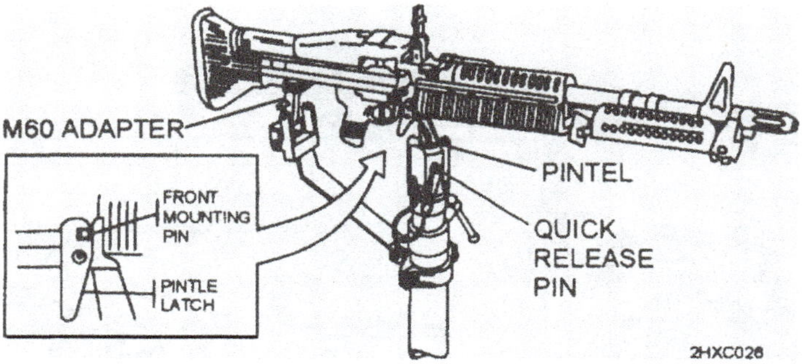

3 Place the end of the machine gun's front mounting pin on top of the pintle assembly.

4 Pivot the M60 adapter assembly into the upright position.

5 Slide the adapter assembly on the machine gun's mounting plate from the rear to the front.

0007-00-4

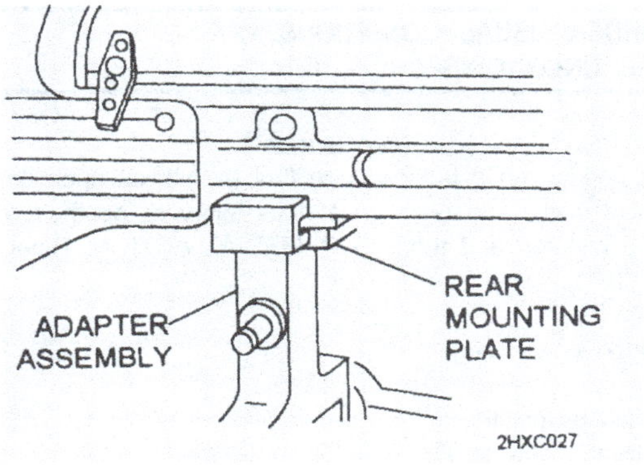

OPERATION UNDER USUAL CONDITIONS - HOW TO PUT M60D ON VEHICLES

0008 00

1 Pull quick release pin.

2 Lower gun adapter.

3 Attach M60D machine gun to aircraft mount using applicable manual.

4 Raise gun adapter.

5 Insert quick release pin

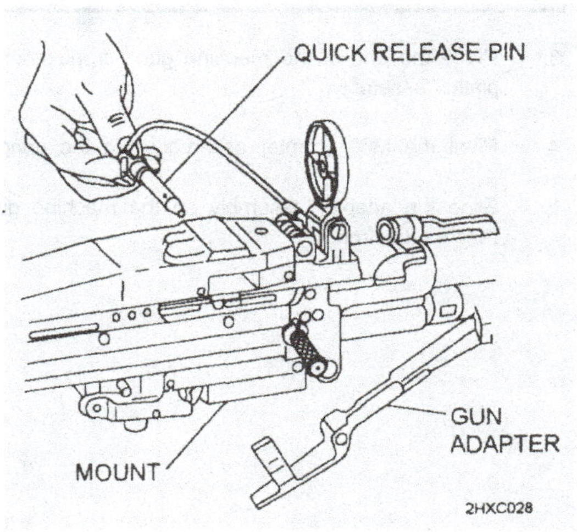

OPERATION UNDER USUAL CONDITIONS - CLEARING AND UNLOADING

0009 00

WARNING

If cover is opened on a hot cartridge (hot barrel), an open cover cookoff could occur and result in serious injury or death. Evacuate area for 15 minutes and then do REMEDIAL ACTION. Refer to WP 0011 00.

1 Point weapon down range.

2 Safety should be on ‘F’.
- M60 - Move safety to ‘F’.
- M60D - Press in on ‘F’.

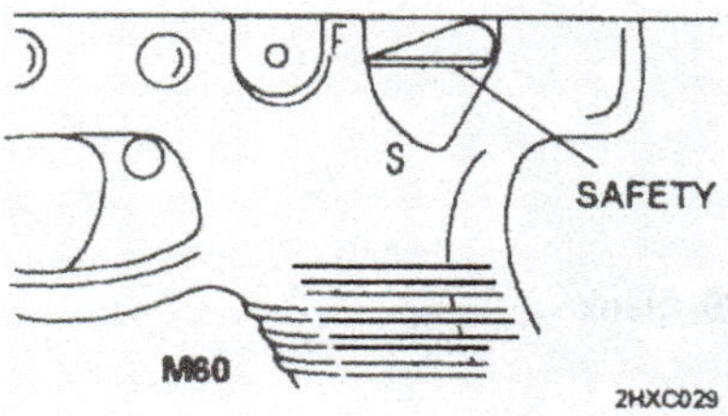

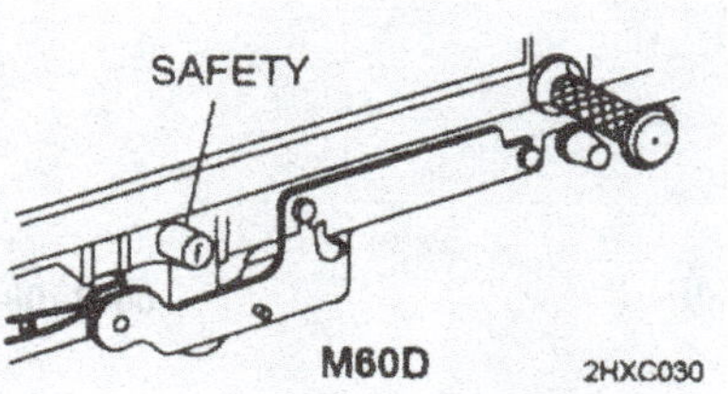

3 Pull cocking handle rearward, then push it forward until it locks.

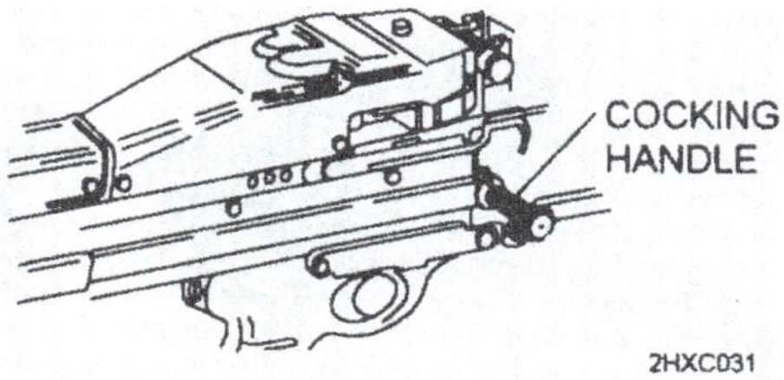

4 M60 - Move safety to 'S'.
M60D - Press in on 'S'.

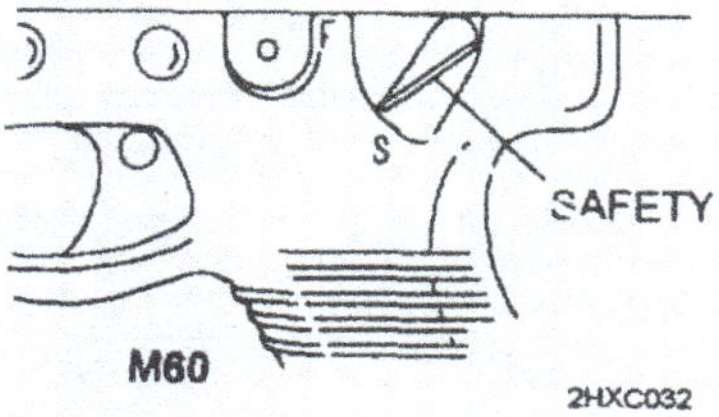

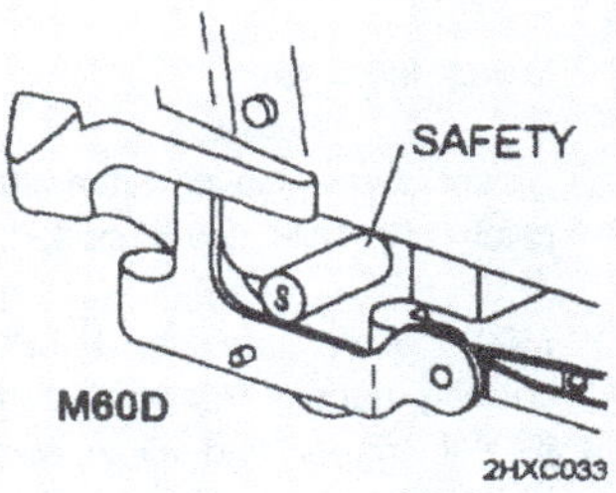

5 Turn latch lever and open cover.

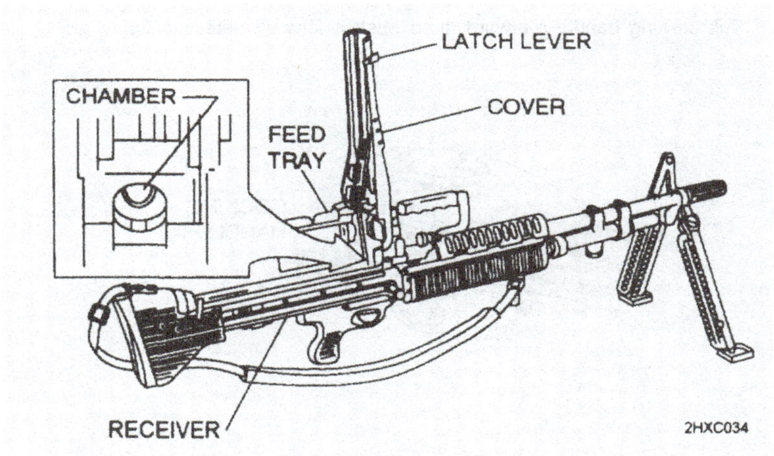

6 Remove ammunition and link belt and raise feed tray.

7 Eyeball chamber. If empty, go to step 8. If a round is chambered, see WP 0012 00 for stuck or ruptured cartridge case and WP 0017 00 for troubleshooting.

8 Lower feed tray.

9 Close cover. Cover latch should catch and hold cover closed.

10 M60 - Move safety to 'F'. Pull cocking handle rearward and hold it. Pull trigger and allow cocking handle to move forward slowly.

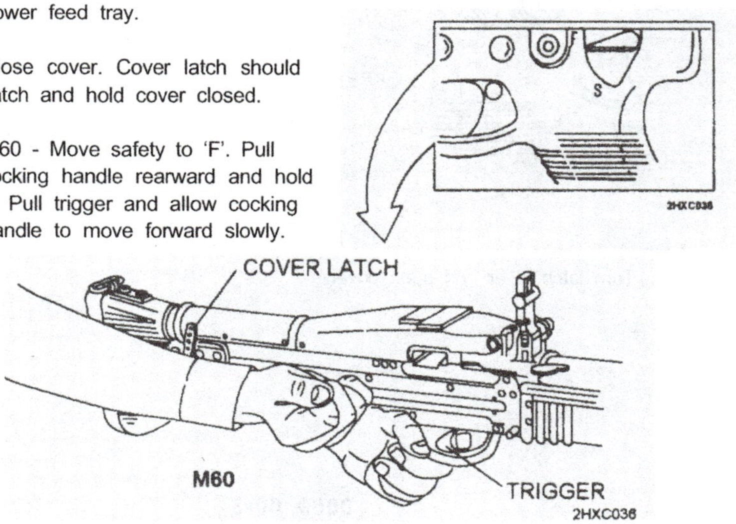

M60D - Press in on 'F'. Pull cocking handle rearward and hold it. Pull trigger and allow cocking handle to move forward slowly.

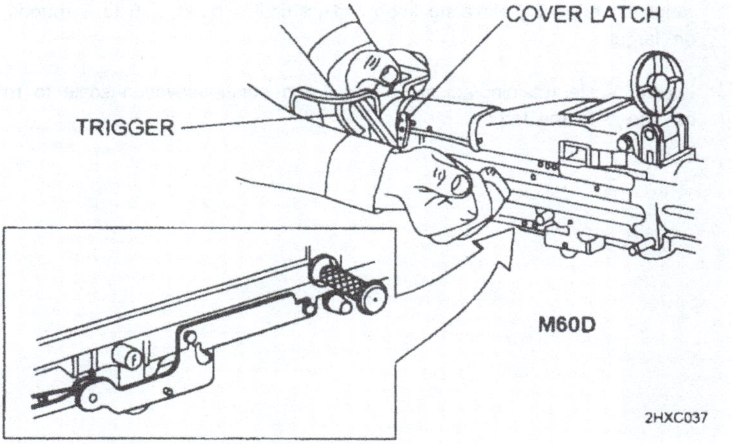

INITIAL ADJUSTMENTS, DAILY CHECKS, AND SELF-TEST 0010 00

FIELD ZEROING

NOTE

One mil is equal to one meter at 1000 meters.

1. Choose a target between 300 and 700 meters.
2. Place range scale to target on rear sights.
3. Zero windage scale with index line.
4. Aim at 6 o'clock on bottom of target.
5. Fire a burst of 6 to 9 rounds and observe impact of burst.

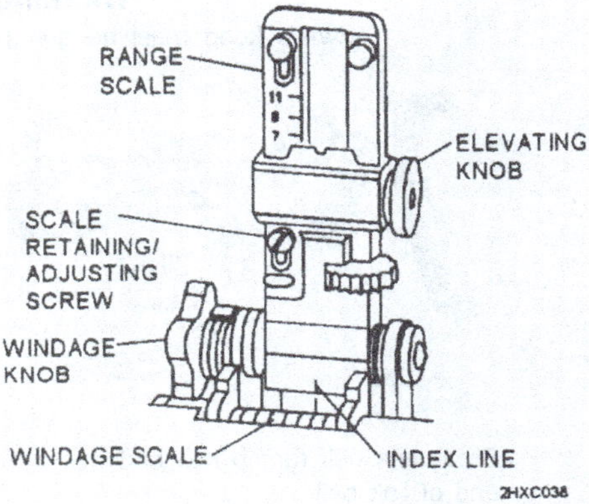

FIELD ZEROING (cont)

6 Correct for left or right wind by adjusting windage knob. Then correct range by adjusting elevating knob. Adjust until a burst of 6 to 9 rounds is on target.

7 Loosen scale retaining/adjusting screw and adjust elevation scale to reflect the range to the target.

IMMEDIATE ACTION

RUNAWAY M60 MACHINE GUN

1 If runaway occurs (machine gun won't stop firing), take the following actions to correct it quickly.

WARNING

Always keep machine gun pointed down range.

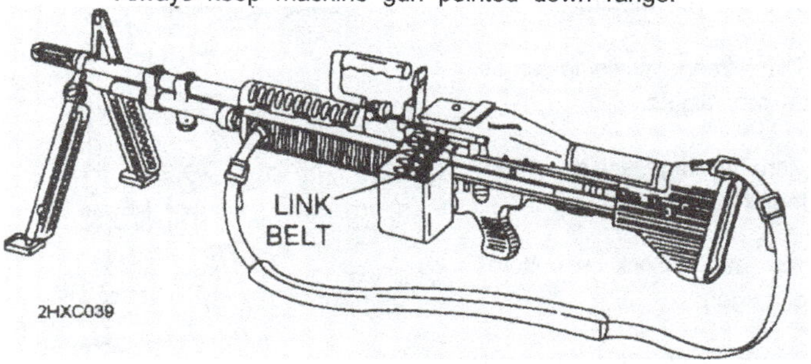

2 Break link belt (grasp belt and twist it firmly) or let machine gun fire if near end of link belt.

IMMEDIATE ACTION (cont)

RUNAWAY M60 MACHINE GUN (cont)

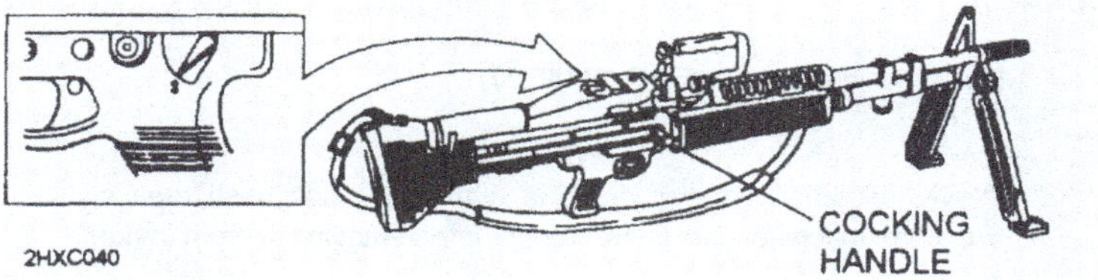

3 Pull cocking handle all the way back and hold it. Place safety to 'S' and remove link belt.

4 Clear machine gun (see WP 0009 00).

WARNING

Never reload a runaway machine gun until it has been repaired.
Be sure machine gun is cleared before moving it.

5 Notify unit maintenance for repair.

RUNAWAY M60D MACHINE GUN

1 If runaway occurs (machine gun won't stop firing), take the following actions to correct it quickly.

WARNING

Always keep machine gun pointed down range.

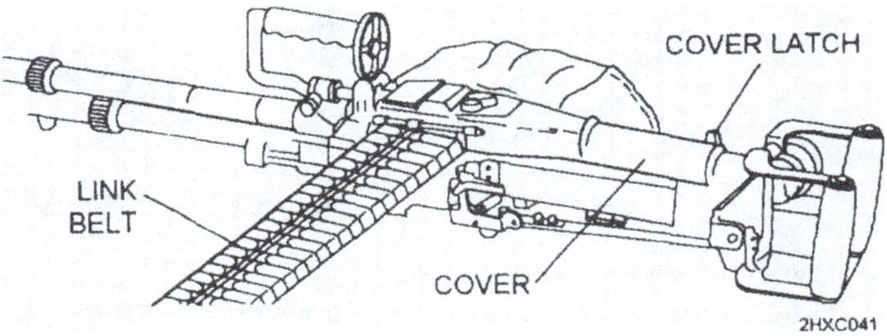

2 Let machine gun fire if near end of link belt.

3 Rotate cover latch and open cover.

IMMEDIATE ACTION (cont)

RUNAWAY M60 MACHINE GUN (cont)

4 Clear machine gun (see WP 0009 00).

WARNING

Never reload a runaway machine gun until it has been repaired. Be sure machine gun is cleared before removing it from mount.

5 Notify unit maintenance for repairs.

MISFIRE

If the weapon stops firing before you complete your mission, take IMMEDIATE ACTION within 10 seconds.

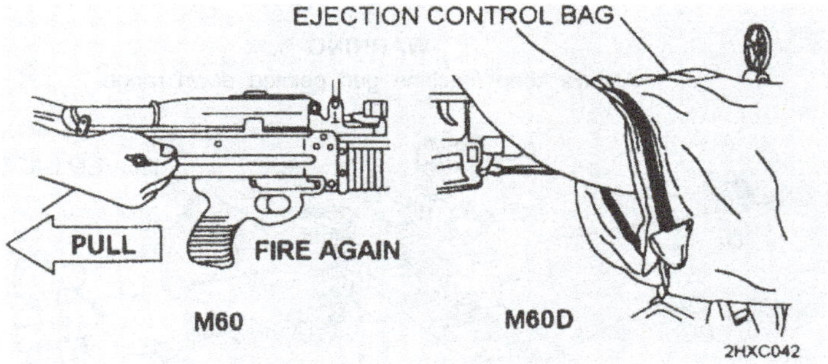

1 M60 - Pull cocking handle rearward.
M60D - Unzip ejection control bag and pull cocking handle rearward.

2 Ensure bolt is locked rearward and return cocking handle to forward locked position.

3 Check ejection port (M60 only).

MISFIRE (cont)

4 If a round or cartridge case is ejected, attempt to fire again.

WARNING

If cover is opened on a hot cartridge (hot barrel), an open-cover cookoff could occur and result in serious injury or death. Close cover and evacuate area for 15 minutes.

The climate temperature in different regions will make a difference as to what constitutes a hot gun. A hot, sunny day can cause a cookoff within 50 rounds, weapon and ammunition in sun.

5 If nothing is ejected and the barrel is hot enough to cause a cookoff (200 rounds fired within 2 minutes), watt at least 15 minutes (make sure bolt is locked rearward). Repeat steps for IMMEDIATE ACTION.

6 If IMMEDIATE ACTION fails to remove cartridge case, take REMEDIAL ACTION.

REMEDIAL ACTION

STEPS TO TAKE IF IMMEDIATE ACTION - MISFIRE - DOES NOT WORK

WARNING

If cover is opened on a hot cartridge (hot barrel), an open-cover cookoff could occur and result in serious injury or death. Close cover and evacuate area for 15 minutes.

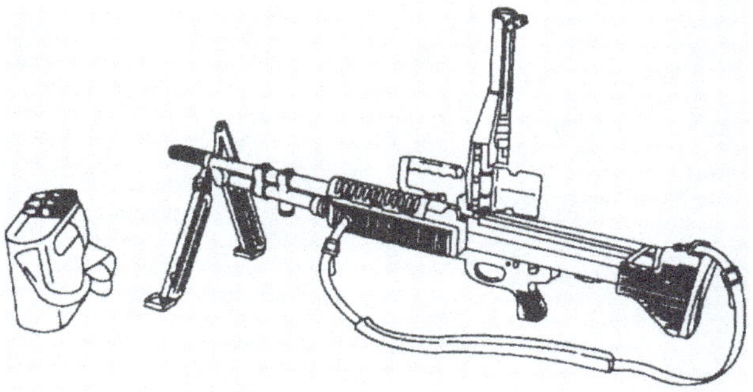

2HXC043

REMEDIAL ACTION (cont)

1 Keep weapon on target (down range/impact area). Clear weapon when barrel is cool (after 15 minutes wait) using the following steps.

2 Pull cocking handle rearward locking bolt to the rear. Return charging handle to forward locked position.

3 If a round is not ejected:
M60 - Place safety to 'S'.
M60D - Press in on 'S'.

4 Open cover, remove ammunition link belt and raise feed tray.

5 Inspect receiver, chamber, ejector, and ammunition.

6 If a round is in chamber, lower feed tray and close cover.

7 M60 - Place safety to 'F'.
M60D - Press in on 'F'.

6 Attempt to fire.

9 If a round is fired and ejected, reload and continue to fire.

10 If weapon does not fire or eject, clear and unload weapon, and notify unit maintenance for repair.

WARNING

When removing a stuck unfired cartridge, stay clear of the muzzle. Do not allow cartridge to contact any hard surface. Cartridge may fire on contact. Remove a stuck unfired cartridge using the same procedures for removing a stuck or ruptured cartridge case.

1 Pull cocking handle rearward and place safety to 'S'.

2 Remove barrel (see WP 0018 00).

3 **STUCK CARTRIDGE CASE/ UNFIRED CARTRIDGE-** Insert cleaning rod in muzzle and tap out case/cartridge.

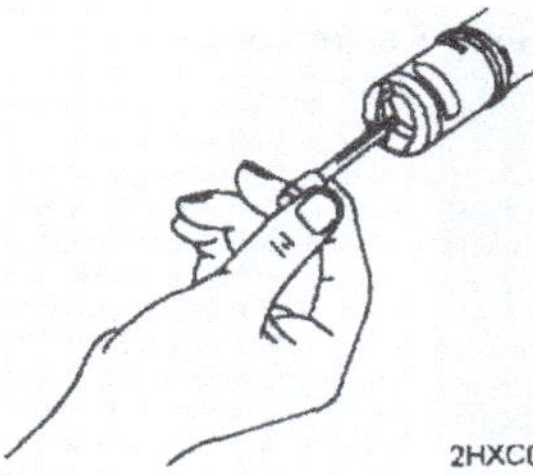

2HXC044

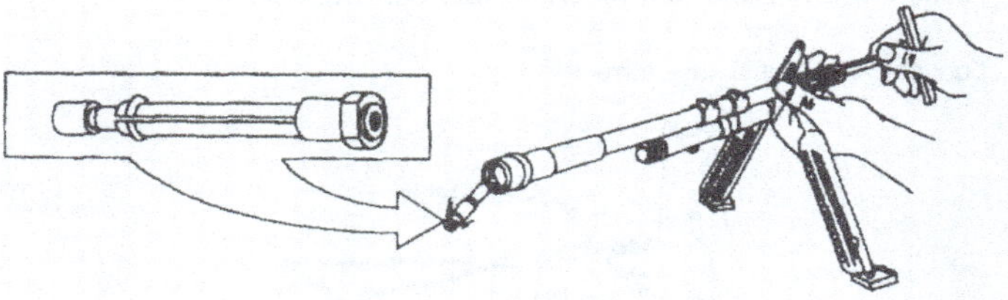

2HXC045

3 **RUPTURED CARTRIDGE CASE (cont)**

a. Push ruptured cartridge extractor inside ruptured case in chamber until it seats.

b. Lightly tap out with cleaning rod.

c. Remove extractor from cartridge case.

d. Unscrew extractor tool and remove cartridge case. Reassemble extractor tool.

4 Install barrel (see WP 0018 00).

5 Close cover.

6 Hold cocking handle rearward.

7 Place safety to 'F'.

8 Pull trigger and ease cocking handle forward.

9 Once weapon is cleared, do not place safety to 'S' when bolt is forward. Leave safety at 'F'.

INSTALLING M13A1 BLANK FIRING ATTACHMENT

1 Loosen lock nut and turn restrictor tube out a few times.

2 Loosen wing nut a few turns.

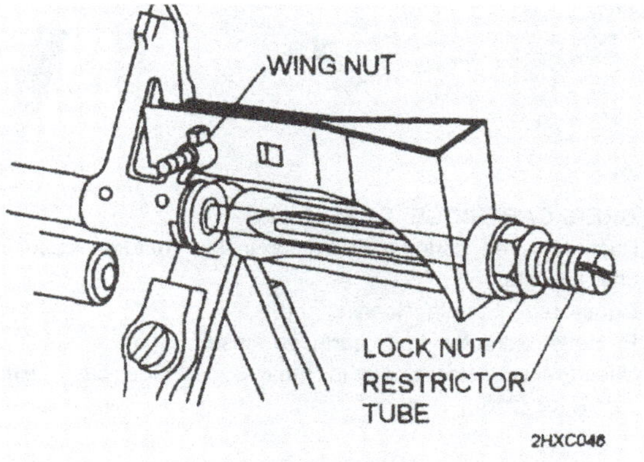

INSTALLING M13A1 BLANK FIRING ATTACHMENT (cont)

3 Slide restrictor tube in flash suppressor as far as possible,

4 Clamp blank firing attachment around front sight.

5 Tighten wing nut fingertight.

6 Screw restrictor tube in until it seats tight against muzzle end of barrel to prevent gas leakage.

7 Tighten lock nut.

NOTE

After Firing 500 Blank Rounds: Clean gun including barrel assembly, gas cylinder, gas piston, gas port, chamber bore, and Blank Firing Attachment (BFA).

NIGHT VISION SIGHT MOUNTED ON M60 MACHINE GUN

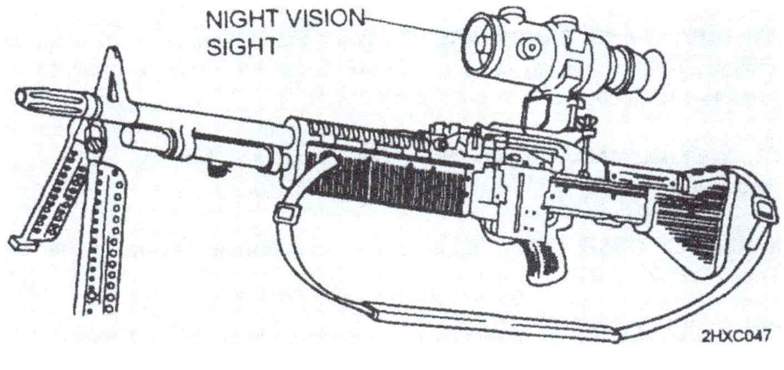

To mount night vision sight AN/PVS-4 on the M60 machine gun, refer to TM 11-5655-213-10.

PREPARATION FOR MOVEMENT 0014 00

Refer to FM 23-67.

OPERATION UNDER UNUSUAL CONDITIONS 0015 00

Under unusual conditions, clean and lubricate machine gun more often.

HOT, DUSTY, AND SANDY AREAS. Clean often. Wipe oil from exposed surfaces with clean wiping rag (item 5, WP 0039 00). Cover weapon as much as possible to keep dust and sand out of parts.

HOT, WET CLIMATE. Inspect often. Dry, clean, and lubricate lightly as necessary.

EXTREMELY COLD CLIMATE. Keep free of moisture. Lightly oil with LAW (item 3, WP 0039 00).

AFTER EXPOSURE TO WATER. Disassemble, clean, oil and reassemble as soon as possible. Make sure it's dry.

NUCLEAR, BIOLOGICAL, AND CHEMICAL (NBC) DECONTAMINATION. Decontamination procedures can be found in FM 3-5.

CHAPTER 3

MAINTENANCE INSTRUCTIONS

LUBRICATION INSTRUCTIONS 0016 00

LUBE GUIDE

Under all but the coldest arctic conditions, LSA (item 4, WP 0039 00), and CLP (item 1, WP 0039 00) are the lubricants to use on your machine gun. Remember to remove excessive oil from the bore before firing.

- Lightly Lubricate. A film of lubricant barely visible to the eye.
- Generously Lubricate. Heavy enough so lubricant can be spread with finger.

NOTE

Lubrication instructions are mandatory. Do not mix lubricants on the same machine gun. The weapon must be thoroughly cleaned during change from one lubricant to another. Dry cleaning solvent (available to unit maintenance) is recommended for cleaning during change from one lubricant to another.

LUBE GUIDE (cont)

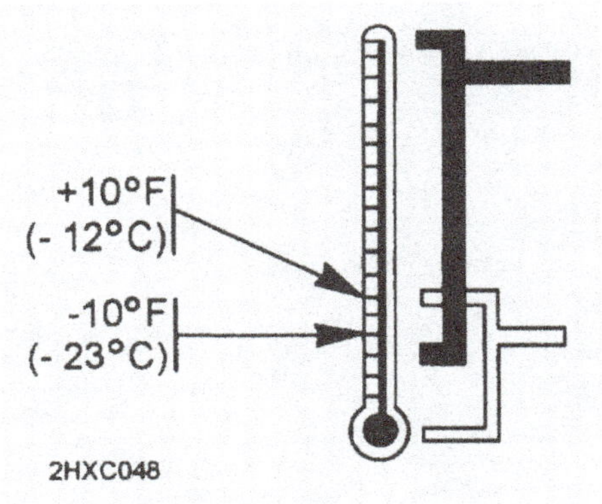

CLP - Cleaner, lubricant and preservative (item 1, WP 0039 00)

LSA - Weapons lubricating oil, semi-fluid (item 4, WP 0039 00)

LAW - Weapons lubricating oil, arctic (item 3, WP 0039 00)

Between 10°F (-12°C) and -10°F (-23°C) use CLP, LSA or LAW.
Below -10°F (-23°C) use only LAW.

WARNING

Be sure to clear weapon before disassembling, cleaning, inspecting, or storing.

1 After firing, field-strip your machine gun (WP 0018 00). Clean bore and chamber with cleaning rod, chamber brush, bore brush, and swab soaked with rifle bore cleaner (RBC) or CLP until a clean swab can be run through the bore without getting dirty.

2 Clean powder-fouled parts, except the buffer, with a wiping rag dampened with RBC or CLP.

3 Wipe dry and lube as required. Inspect and run a lightly oiled swab through the bore and chamber.

4 If your machine gun is not used, it still needs complete cleaning and lubing at least every 90 days. (Unusual conditions could shorten this time.)

LUBE GUIDE (cont)

5 External Surfaces. Put CLP on a clean swab and generously lubricate all shaded areas. Lightly lubricate all other areas except rubber-coated parts.

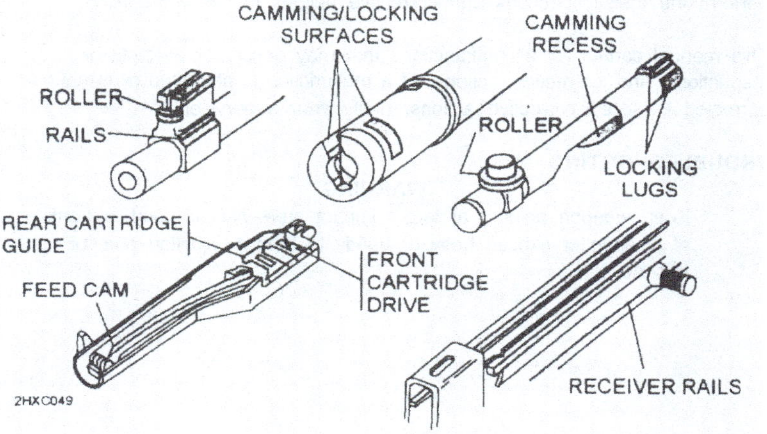

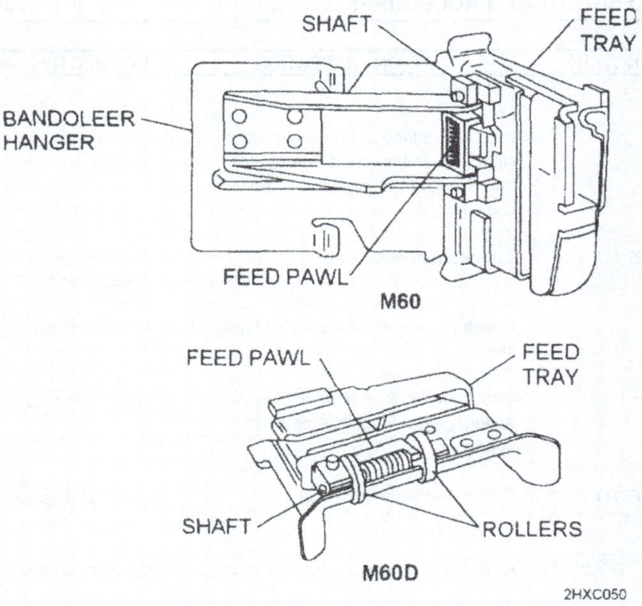

INTRODUCTION

The table lists the common malfunctions which you may find during the operation or maintenance of the machine gun or its components. You should perform the tests/inspections and corrective actions in the order listed.

This manual cannot list all malfunctions that may occur nor all tests or inspections and corrective actions. If a malfunction is not listed or is not corrected by listed corrective actions, notify your supervisor.

TROUBLESHOOTING

WARNING

Keep weapon pointed at target impact area. Never stand in front of weapon or expose body or hands to breech, ejection port, or muzzle.

MALFUNCTION	TEST OR INSPECTION	CORRECTIVE ACTION
COOKOFF*	Hot barrel	
	If you have tired 200 or more rounds within 2 minutes (rapid rate of tire), barrel will be hot enough to cause cookoff.	Cool weapon for 15 minutes.
FAILURE TO FIRE	Check ammunition.	Replace faulty ammunition.
	Check for broken or damaged firing pin.	Notify unit maintenance
	Check for weak or broken firing pin spring, guide assembly, or driving spring	Notify unit maintenance
FAILURE TO EXTRACT	Broken extractor or spring.	Notify unit maintenance

*COOKOFF - A misfire of a chambered round caused by the weapon overheating

MALFUNCTION	TEST OR INSPECTION	CORRECTIVE ACTION
FAILURE TO EXTRACT (cont)	Short recoil.	Clean gas port, piston, gas cylinder, operation rod assembly, and chamber (see WP 0023 00).
	Gas piston installed backwards	Install correctly.
	Dirty ammunition or chamber.	Clean chamber and/or use new ammunition.
SLUGGISH OPERATION	Friction from dirt, carbon, burrs or lack of lubrication.	Clean and lubricate or notify unit maintenance of any burrs.
WADDING AND UNBURNED POWDER	Manually charge weapon (if firing blanks)	Clean and lubricate or notify unit maintenance.
FAILURE TO CHAMBER	Ruptured cartridge case	Remove cartridge case (WP 00012 00).
	Carbon buildup in gas cylinder	Remove carbon.
	Carbon buildup in receiver assembly.	Remove carbon

MALFUNCTION	TEST OR INSPECTION	CORRECTIVE ACTION
FAILURE TO CHAMBER (cont)	Damaged round.	*Remove round and recharge weapon.
	Dirty chamber.	Clean or replace barrel.
FAILURE TO EJECT	Frozen, damaged or missing ejector spring.	Notify unit maintenance.
	Short recoil.	Clean gas port, piston, gas cylinder, chamber, and operating rod guide (WP 0023 00).
FAILURE TO COCK OR RUNAWAY GUN	Operating rod sear notch or sear worn.	Notify unit maintenance
	Sear plunger or spring broken or missing.	Notify unit maintenance.

*Dispose of damaged round according to range SOP

MALFUNCTION	TEST OR INSPECTION	CORRECTIVE ACTION
FAILURE TO FEED	Insufficient gas pressure (short recoil).	Clean gas system, chamber, and barrel socket (WP 0023 00).
	Improper lubrication.	Lubricate as necessary (WP 0016 00).
	Defective ammunition or links.	Use new ammunition belt.
	Ammunition belt installed wrong.	Install with open side of link down (WP 0005 00).
	Damaged or weak operating parts.	Notify unit maintenance
	Obstruction in receiver assembly.	Remove

END OF TASK

FIELD-STRIPPING MACHINE GUN 0018 00

M60 SHOULDER GUN STOCK

WARNING

Before field stripping, clear the weapon and check that the bolt is in forward position.

NOTE

The following procedures illustrate the M60 machine gun except where M60 parts differ from M60D pans, then both weapons are illustrated.

1 Lift shoulder rest (1), insert cleaning rod (2) in hole to release latch. Pull shoulder gun stock (3) from receiver.

2 Turn latch lever (4) and open cover.

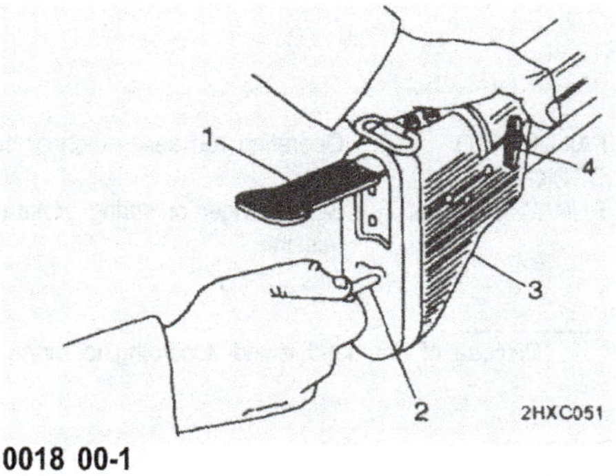

M60D SPADE AND TRIGGER GRIP ASSEMBLY

WARNING

To prevent possible bodily injury and aircraft damage, personnel should not stand the weapon on its barrel assembly when disassembling or assembling the weapon.

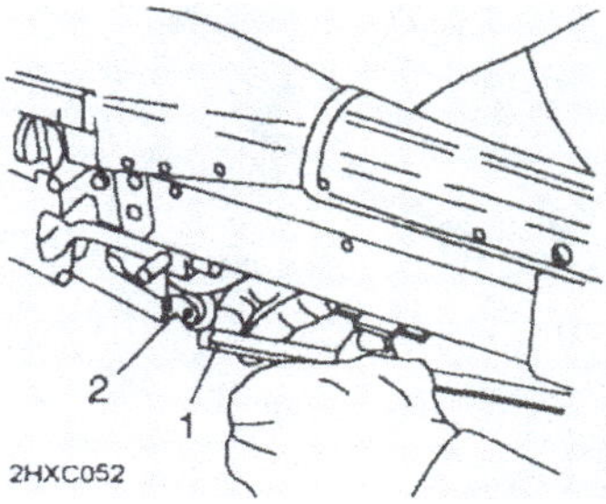

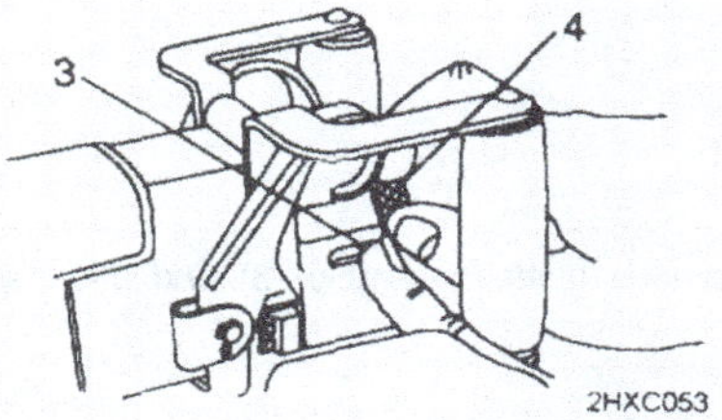

1 Unsnap sear assembly link and spring (1) from rear of sear assembly activator (2) by pressing upward with fingers.

2 Press and hold end of spring pin (3) forward while unscrewing knob and pin assembly (4).

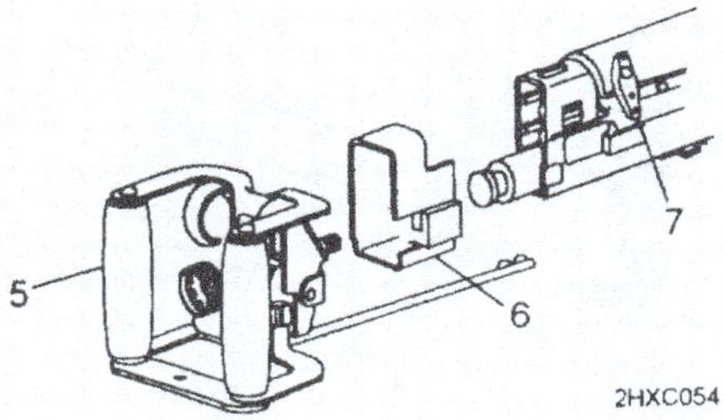

3 Remove grip and trigger assembly (5) and dust and moisture boot (6).

4 Turn latch lever (7) and open cover.

BOLT ASSEMBLY AND OPERATING ROD ASSEMBLY

1 Apply slight palm pressure to rear of hydraulic buffer assembly (1). Lift out buffer retaining yoke (2) and remove buffer assembly (1).

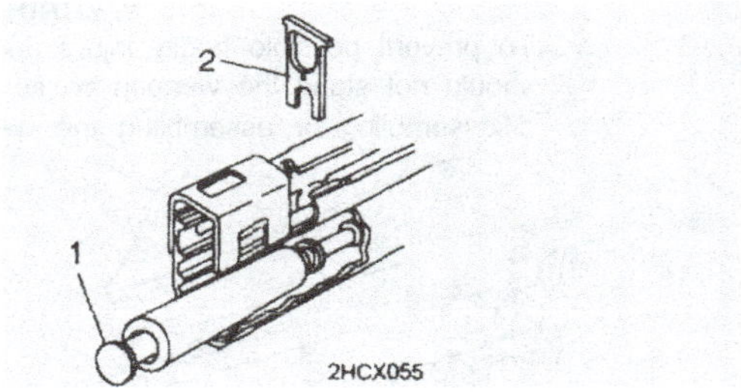

2 Remove guide assembly (3) and drive spring (4).

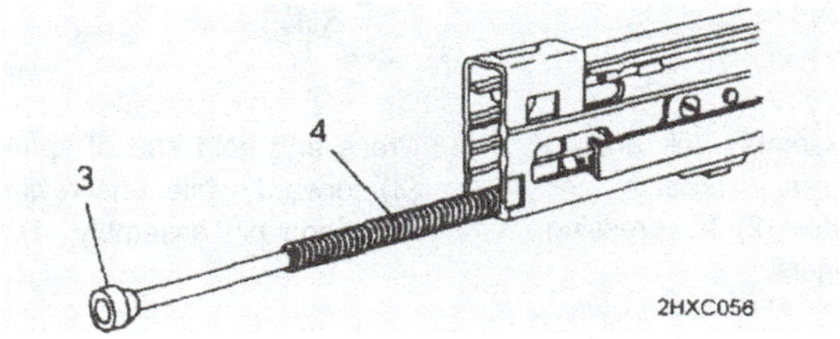

3 M60. Place safety to 'F'. M60D. Press in on 'F'.

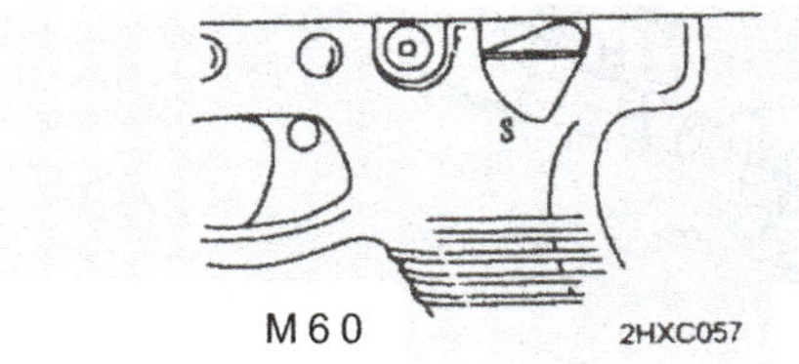

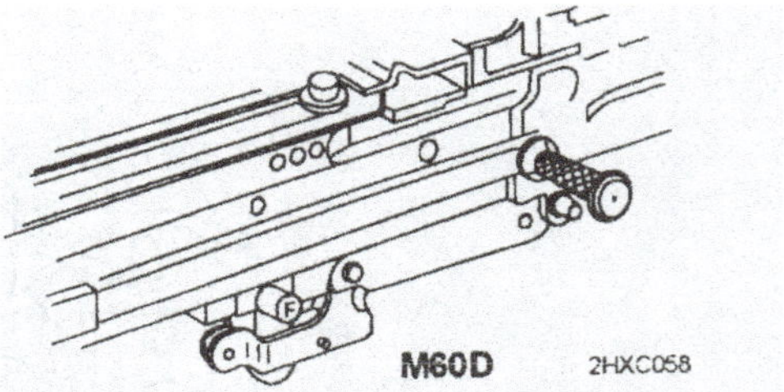

4 M60. Pull trigger (5) and hold it while pulling cocking handle assembly (6) rearward.

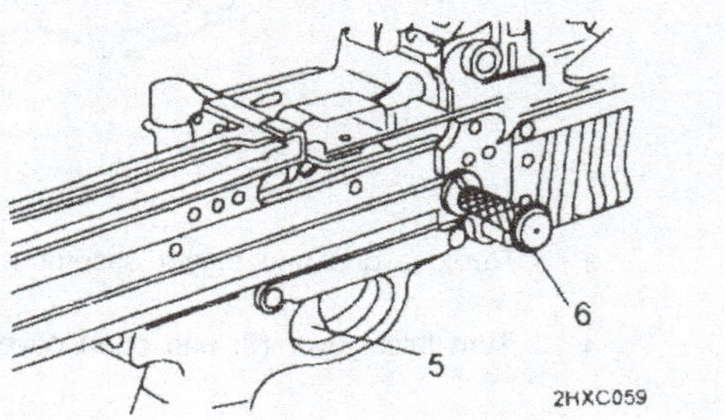

BOLT ASSEMBLY AND OPERATING ROD ASSEMBLY (cont)

M60 D. Pull down and hold sear assembly activator (7) while pulling cocking handle assembly (6) rearward.

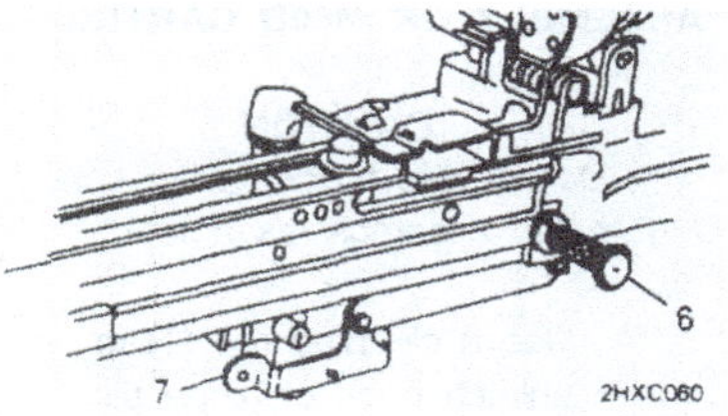

WARNING

Bolt assembly is under spring tension; it can twist and injure your hand.

5 Removing operating rod assembly (8) and bolt assembly (9) as a unit.

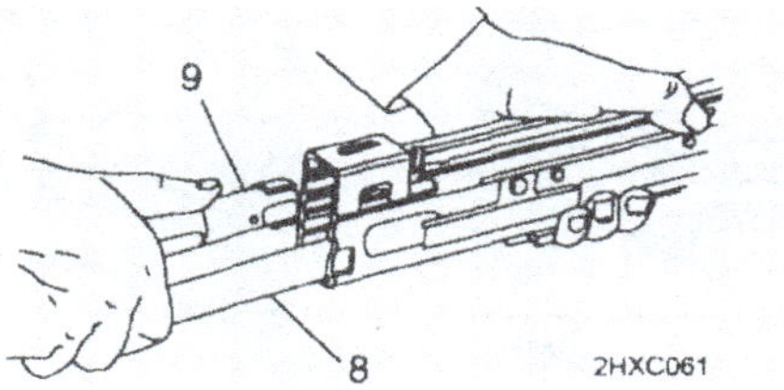

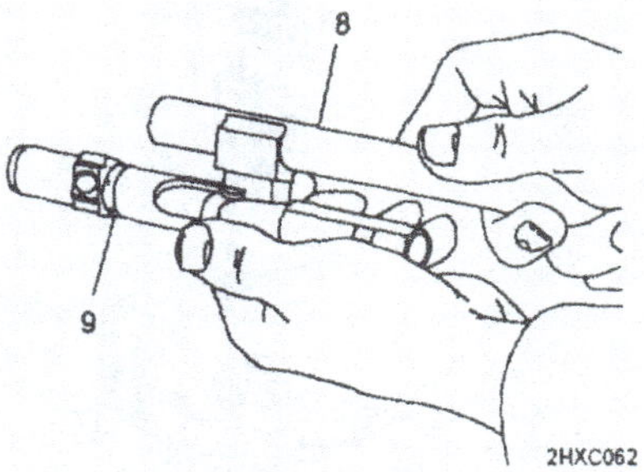

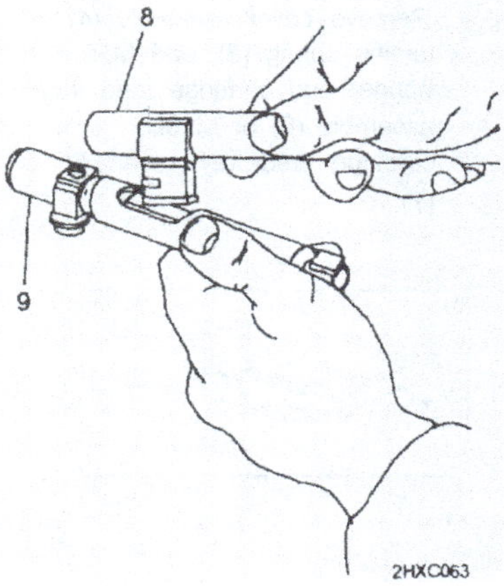

6 Push operating rod assembly (8) toward rear of slot in bolt assembly (9).

7 Separate operating rod assembly (8) from bolt assembly (9).

COVER ASSEMBLY AND M60 HANGER AND CARTRIDGE FEED TRAY ASSEMBLY OR M60D CARTRIDGE FEED TRAY ASSEMBLY

CAUTION

Do not use tip of driving spring guide assembly as a tool.

Use a cleaning rod (1) to unlatch hook of hinge pin latch (2). Remove hinge pin latch (2) and cover hinge pin (3).

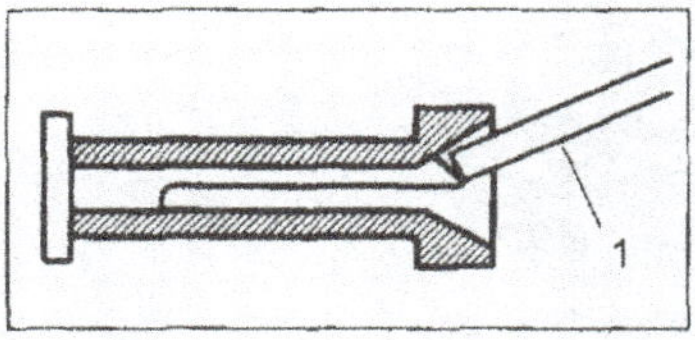

2HXC064

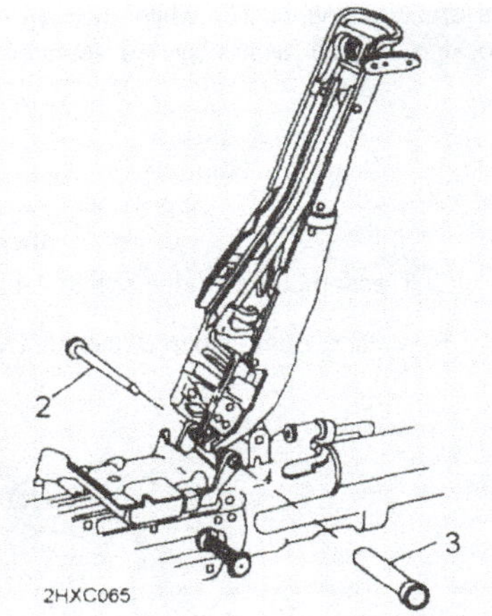

2HXC065

2 Remove cover assembly (4), torsion spring (5), and M60 hanger and cartridge feed tray assembly (6) or M60D cartridge feed tray assembly (7).

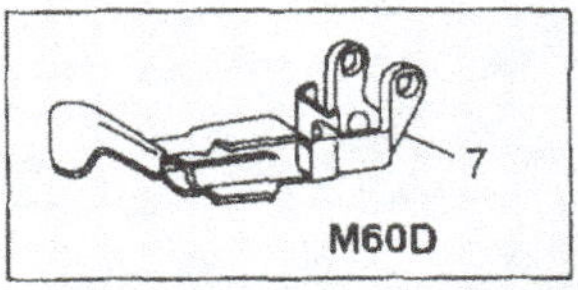

2HXC066

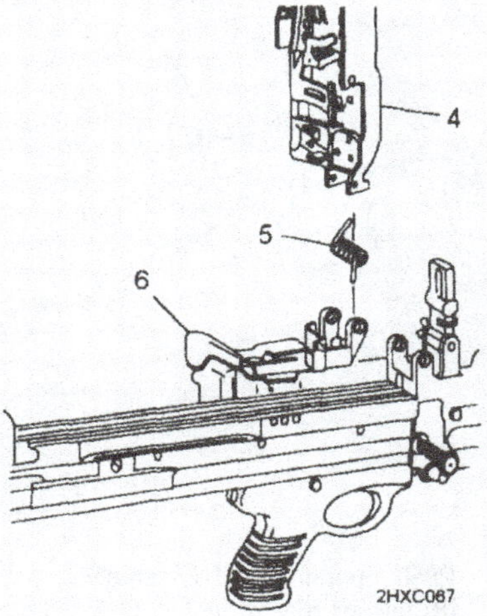

2HXC067

BARREL AND BIPOD ASSEMBLY

1 M60D. Pull quick release pin (4) out of receiver assembly and let gun adapter (5) swing down.

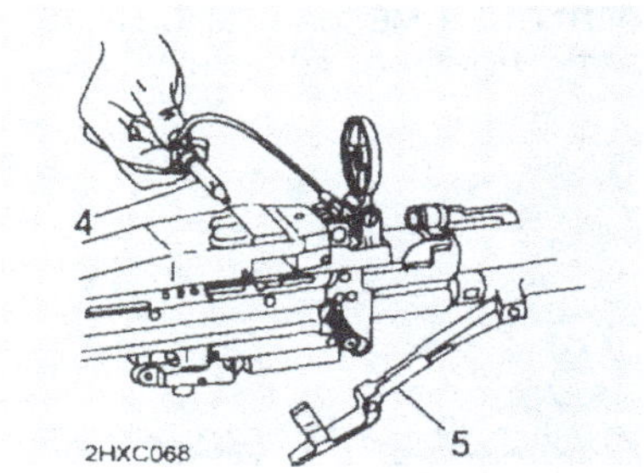

2 Push in spring detent (1) and raise barrel lock (2). Remove barrel and bipod assembly (3).

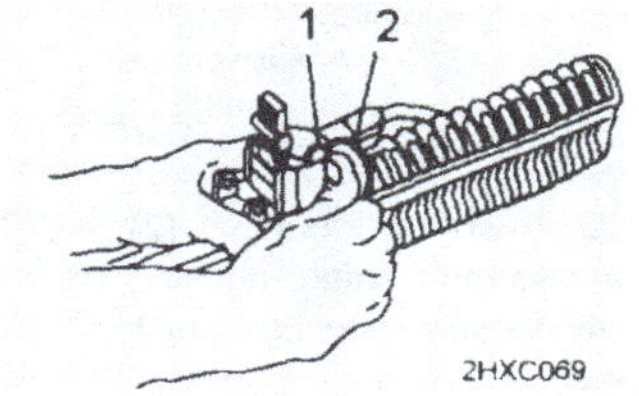

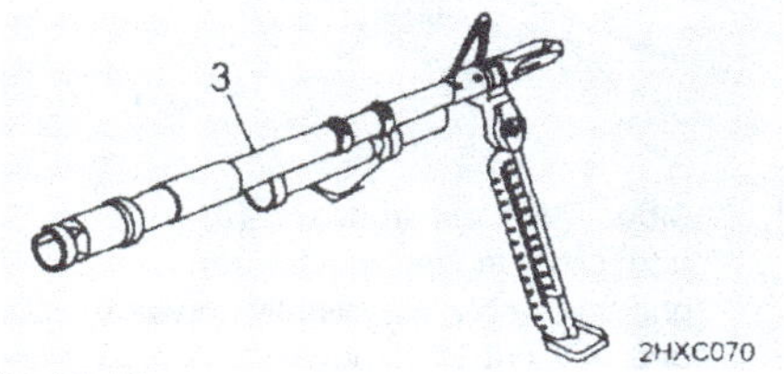

M60 TRIGGER MECHANISM GRIP ASSEMBLY OR M60D SEAR AND SAFETY HOUSING ASSEMBLY

CAUTION

Care should be taken, when disassembling M60 and M60D in the same work area, to ensure proper leaf spring remains with the appropriate model weapon.

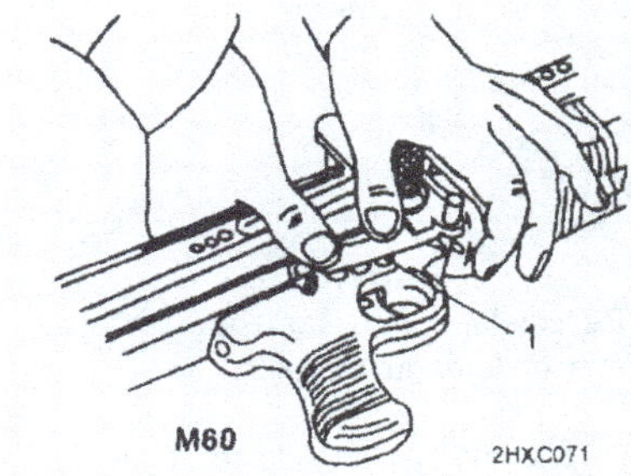

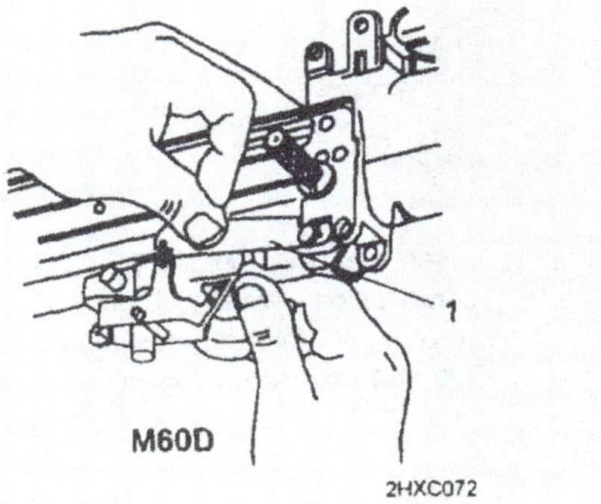

M60 and M60D. Push in and remove flat leaf spring (1).

M60 TRIGGER MECHANISM GRIP ASSEMBLY OR M60D SEAR AND SAFETY HOUSING ASSEMBLY (cont)

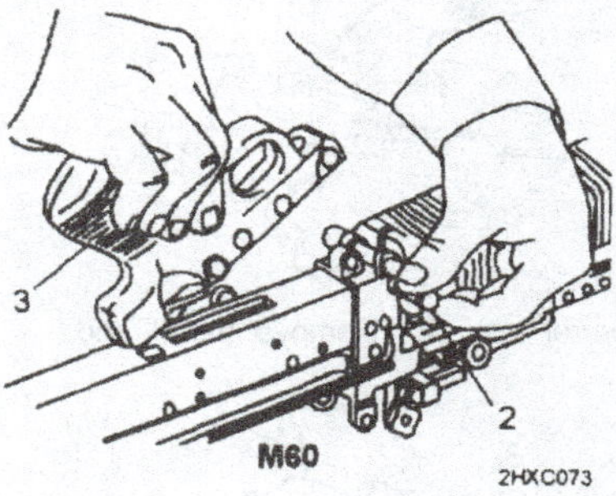

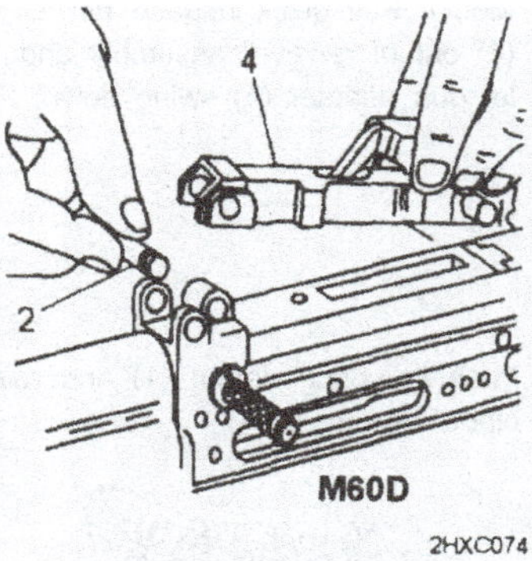

2 M60. Push out front pin (2) and slide trigger mechanism grip assembly (3) slightly forward and pull out to remove.

M60D. Push out front pin (2) and slide sear and safety housing (4) slightly forward and pull out to remove.

FOREARM ASSEMBLY

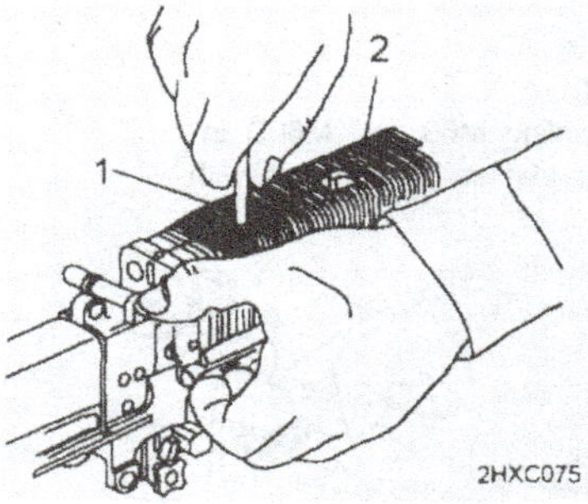

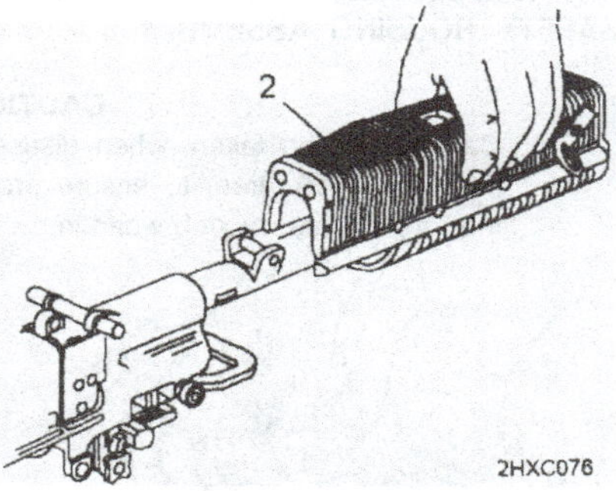

1 Insert cleaning rod (1) or combination wrench through opening in forearm assembly (2) and push down on spring.

CAUTION

Be careful not to damage internal ribs of forearm assembly.

2 Lift and gently slide forearm assembly (2) from receiver.

END OF TASK

INSPECT AND CLEAN

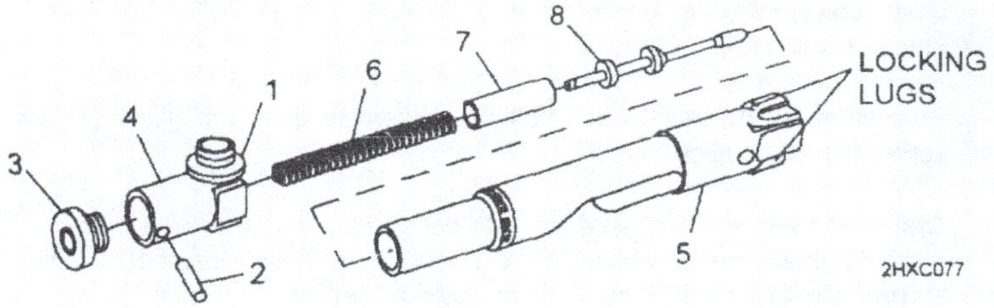

1 Inspect bolt assembly for burrs and cracks, especially in the locking lug area.

2 Make sure roller (1) operates freely and is not cracked.

INSPECT AND CLEAN (cont)

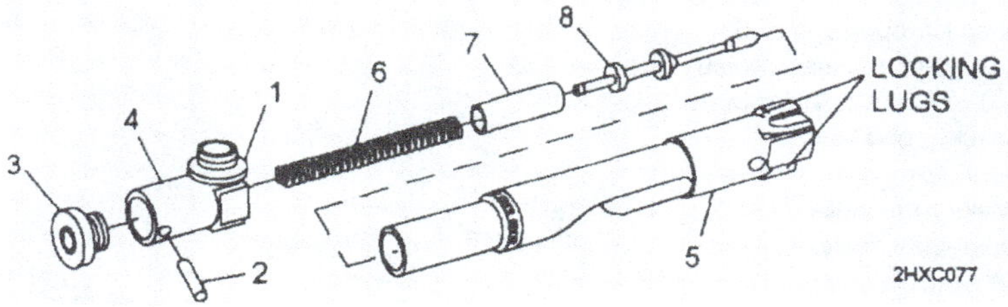

NOTE

Rotate cam actuator assembly so holes line up with headless straight pin. Use reamer of combination wrench to remove or replace headless straight pin.

3 Remove headless straight pin (2) and unscrew bolt plug assembly (3). Remove cam actuator assembly (4) from breech bolt (5). Separate spring (6), guide (7). and firing pin (8).

4 Make sure spring (6) is not kinked.

5 inspect threads on bolt plug assembly (3) and in breech bolt (5) for damage.

6 Clean powder fouling, corrosion, and dirt from all parts with CLP/RBC (items 1 and 2,WP 0039 00).

7 Drop firing pin (8) into breech bolt (5). Assemble and drop guide (7) and spring (6) into breech bolt (5).

8 Assemble cam actuator assembly (4) on breech bolt with roller to front. Screw in bolt plug assembly (3) and align three holes. Insert headless straight pin (2). Make sure cam actuator assembly (4) rotates freely when headless straight pin (2) is fully seated.

END **OF** TASK

MAINTENANCE OF OPERATING ROD ASSEMBLY 0020 00

INSPECT AND CLEAN

1 Inspect operating rod (1), yoke (2), sear notch (3), and pins (4) for burrs, cracks, and chips.

2 Make sure roller (5) operates freely.

3 Inspect driving spring (6) for tension, kinks, breaks, and wear.

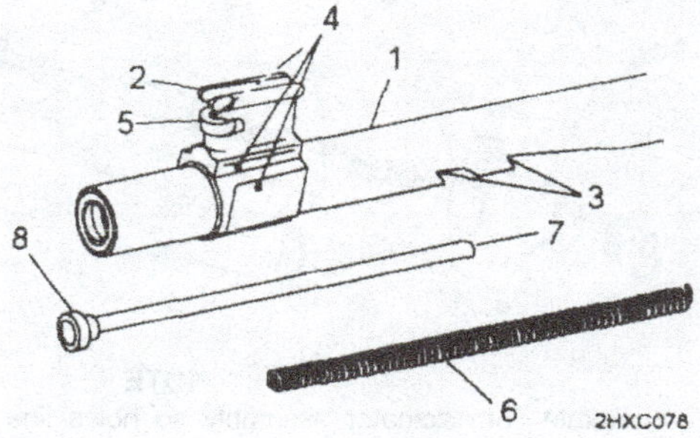

4 Check guide assembly (7) to make sure stop (8) is tight. If loose, notify unit maintenance.

5 Clean powder fouling, corrosion, and dirt from all parts with CLP/RBC (items land 2, WP 0039 00).

INSPECT AND CLEAN (cont)

6 Clean fluted carbon remover (9),

NOTE

Use fluted carbon remover (9) to clean inside of receiver (see WP 0025 00).

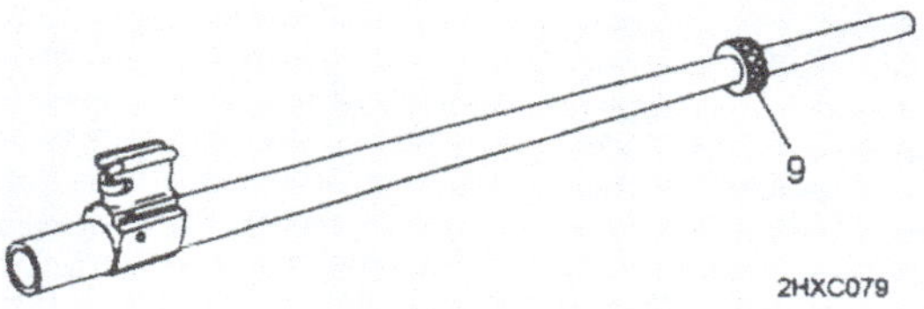

END OF TASK

MAINTENANCE OF COVER ASSEMBLY (M60 AND M60D) 0021 00

INSPECT AND CLEAN

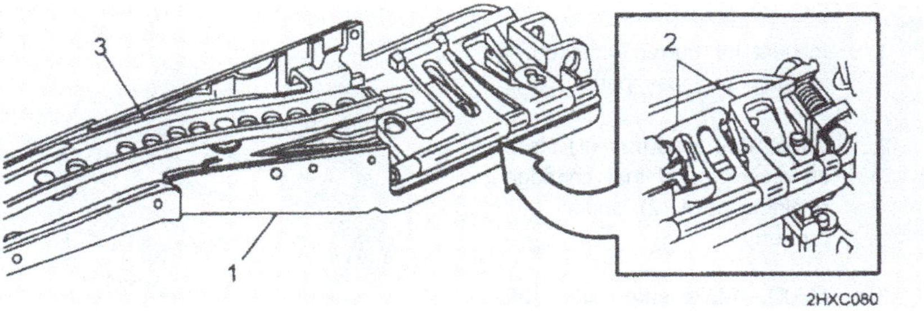

1 Inspect cover assembly (1) for bent, burred, cracked, loose, or missing parts.

2 Check spring action of front and rear cartridge guides (2), feed pawls (beneath the cartridge guides), and feed cam assembly (3).

INSPECT AND CLEAN (cont)

3 Make sure feed cam assembly is secure and operates freely.

4 Clean powder fouling, corrosion, and dirt from all areas with CLP/RBC (items 1 and 2, WP 0039 00).

END OF TASK

MAINTENANCE OF M60 HANGER AND CARTRIDGE FEED TRAY ASSEMBLY OR M60D CARTRIDGE FEED TRAY ASSEMBLY 0022 00

INSPECT AND CLEAN

1 Inspect for burred, cracked, missing, or loose parts.

2 Check that feed tray (1) fits on receiver and that cartridge retaining pawl (2) works properly.

3 M60D - Make sure rollers (3) do not bind.

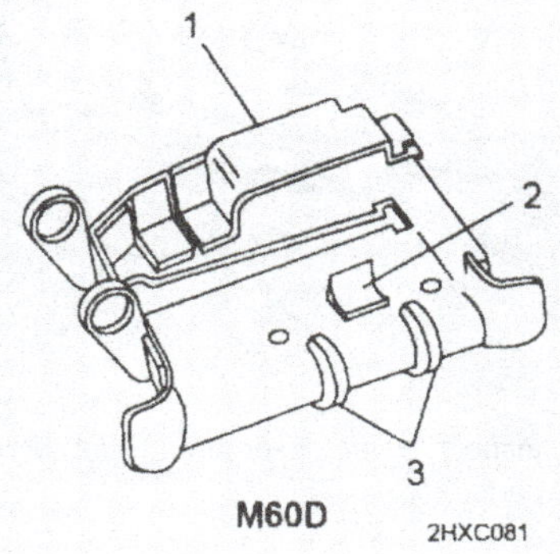

INSPECT AND CLEAN (cont)

4 M60 - Inspect for obstructions in hanger (4).

5 Clean corrosion and dirt from ail areas with CLP/RBC (items 1 and 2, WP 0039 00).

6 Remove CLP/RBC from top surface of feed tray (1) if gun is to be fired.

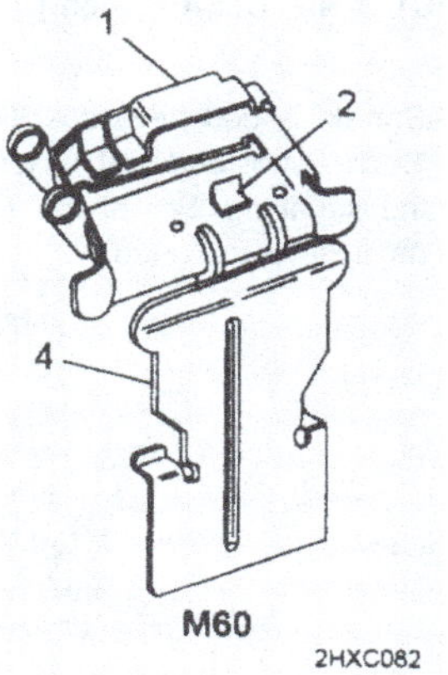

END OF TASK

MAINTENANCE OF BARREL AND BIPOD ASSEMBLY (INCLUDING GAS SYSTEM) 0023 00

BIPOD ASSEMBLY

INSPECT AND CLEAN

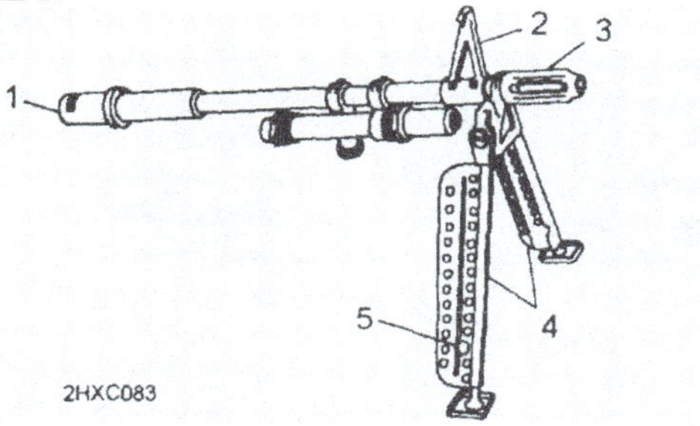

1 Check for burrs, cracks, and wear especially in the barrel socket area (1).

2 Make sure sight (2) and flash suppressor (3) are tight.

3 Make sure bipod (4) works, its legs are straight, and connections are tight.

4 Make sure bipod plungers (5) operate smoothly.

BIPOD ASSEMBLY (cont)

INSPECT AND CLEAN (cont)

WARNING

Barrels issued for a specific gun will not be changed from gun to gun. Each barrel and bolt assembly should be tagged during cleaning with the serial number of the receiver (gun) and should remain together as initially assigned.

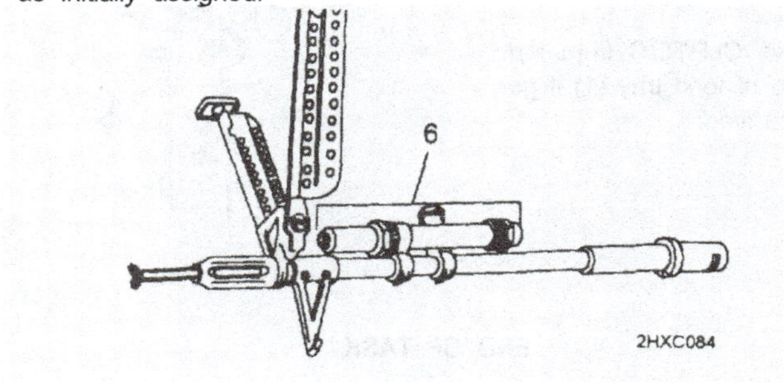

0023 00-2

5 Turn barrel upside down for cleaning. This will keep gas piston and inside of gas cylinder (6) dry.

6 Run a cleaning rod with swab (item, 6 WP 0039 00) saturated with CLP/ RBC (items 1 and 2,WP 0039 00) through bore. Leave bore wet.

GAS SYSTEM

INSPECT AND CLEAN

CAUTION

No abrasive materials will be used to clean the gas piston or inside the gas clinder.

NOTE

Gas piston clicks when barrel is tilted. Take gas system (gas cylinder) apart only when you don't hear a click.

1 Use combination wrench and clean swab (item 6, WP 0039 00) to clean powder fouling, corrosion, and dirt from all parts.

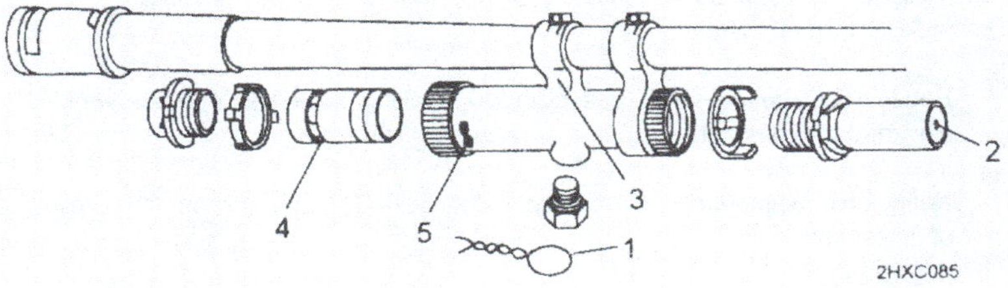

2 Remove safety wire (1) (notify unit maintenance).

3 Disassemble gas cylinder with combination wrench

4 Clean extension vent hole (2) with safety wire.

5 Clean gas port (3), gas piston hole (4), and gas cylinder vent holes (5) with reamer portion of wrench.

GAS SYSTEM (cont)

INSPECT AND CLEAN (cont)

CAUTION

Insert reamer carefully to prevent damage.

6 Bottom the reamer in the hole in the gas port (3) to make sure you have removed all carbon. Run small arms cleaning brush (bore) through bore to remove carbon deposits.

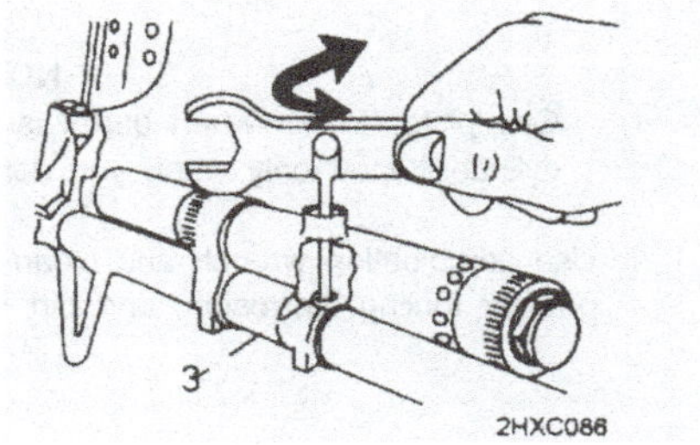

CAUTION

Make sure you do not install piston backward. Do not get CLP/ RBC in gas system.

7 Reassemble gas system (gas cylinder), making sure gas cylinder vent, plug and nut are tight.

8 Notify unit maintenance to replace safety wire (1) promptly on completion of mission.

9 When safety wire is not available, check gas system periodically for tightness.

END OF TASK

INSPECT AND CLEAN

1 Hold down on sear (1) and remove sear pin (2).

2 Remove sear (11, sear plunger (3), and spring (4).

3 Remove trigger pin (5) and trigger (6).

4 Clean powder fouling, corrosion, and dirt from all areas with CLP/RBC (items 1 and 2, WP 0039 00) and wipe with a small arms cleaning brush.

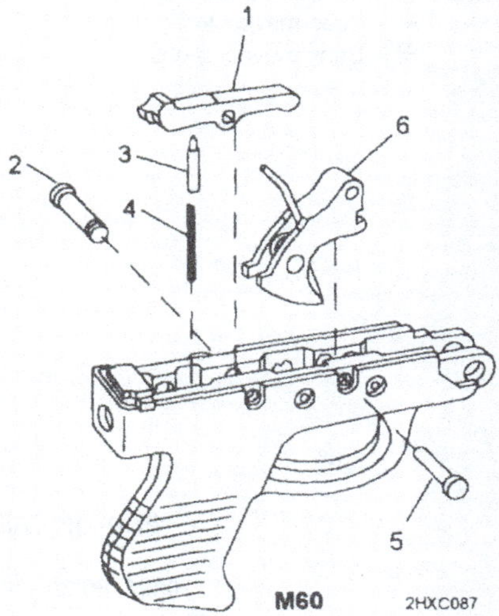

INSPECT AND CLEAN (cont)

5 Inspect sear (1) for chips, cracks and signs of wear.

6 Inspect sear plunger (3) and spring (4) for signs of wear.

7 Position trigger (6) in the grip so that the trigger spring is under the channel. Align holes and insert trigger pin (5) through the right side of grip and secure trigger (6).

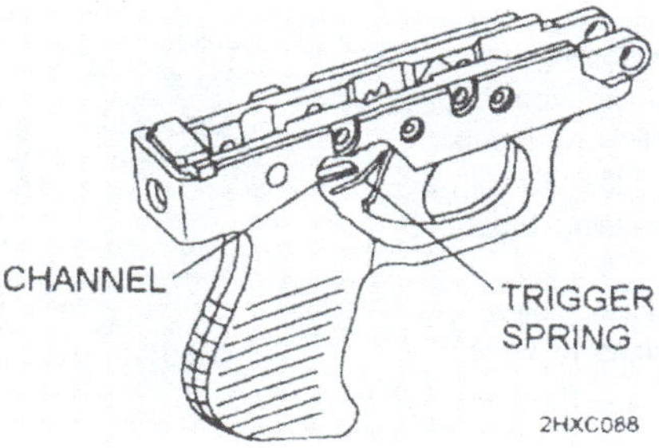

8 Insert spring (4) and sear plunger (3) into hole. Position sear (1) and apply slight pressure to align holes. Insert sear pin (2) through left side of grip and secure sear.

9 Check safety while pulling trigger back:
- 'F' - Sear moves freely.
- 'S' - Sear moves slightly.

END OF TASK

MAINTENANCE OF M60D SEAR AND SAFETY HOUSING ASSEMBLY — 0025 00

INSPECT AND CLEAN

1 Hold down top of sear (1) and remove sear pin (2) sear assembly activator (3) and sleeve bushing (4).

2 Lift out sear (1) sear plunger (5), and spring (6).

3 Clean powder fouling, corrosion, and dirt from all parts with CLP/RBC (items 1 and 2, WP 0039 00) with a small arms cleaning brush.

4 Check sear (1) for chips, cracks, and signs of wear.

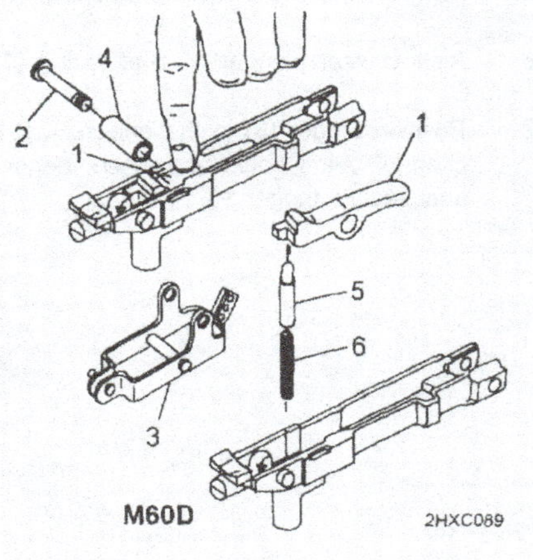

INSPECT AND CLEAN (cont)

5 Check sear plunger (5) and spring (6) for signs of wear.

6 Insert spring (6) and sear plunger (5) into hole. Position sear (1) into housing.

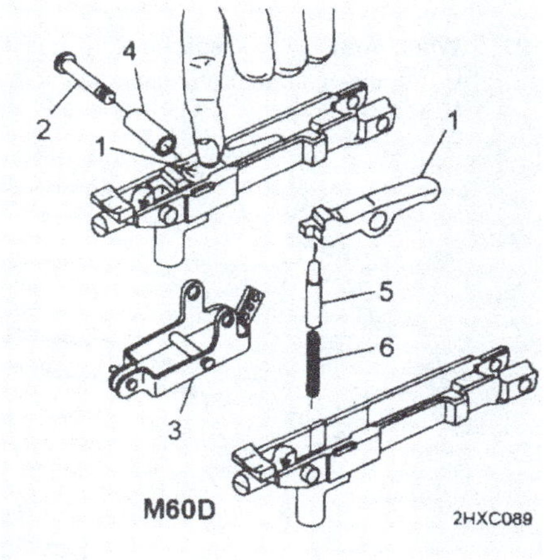

7 Apply slight pressure to align holes and insert sleeve bushing (4).

8 Position sear assembly activator (3) on bottom of sear housing and secure with sear pin (2).

9 Check safety while pulling down on sear assembly activator.
- 'F' - Sear moves freely.
- 'S' - Sear moves slightly.

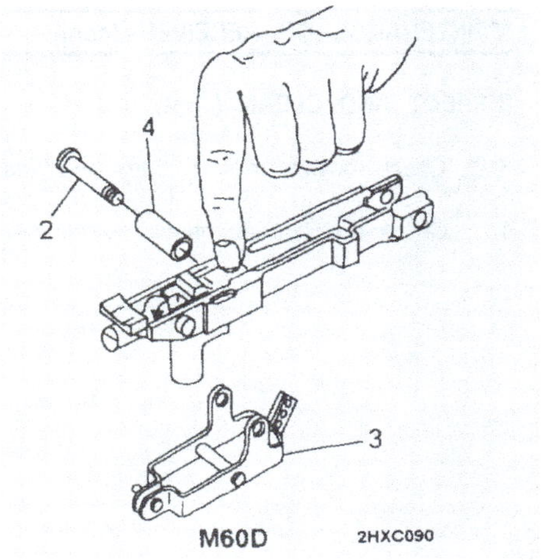

END OF TASK

INSPECT AND CLEAN

1 Clean powder fouling, corrosion, and dirt from all areas with CLP/RBC (items 1 and 2, WP 0039 00) and small arms brushes.

2 When weapon is disassembled, use fluted end of operating rod assembly to remove carbon from operating rod guide of receiver (WP 0020 00).

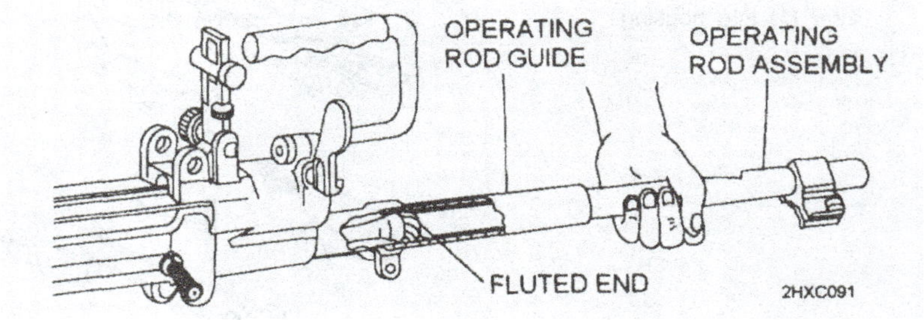

INSPECT AND CLEAN (cont)

3 Check receiver rails for burrs and wear.

4 Check that cocking handle moves freely.

END OF TASK

M60 FOREARM ASSEMBLY

CAUTION

When installing forearm assembly, be careful not to damage its internal ribs.

1 Slightly tip forearm assembly and slide it onto receiver.

2 Press in on bottom of forearm assembly to latch.

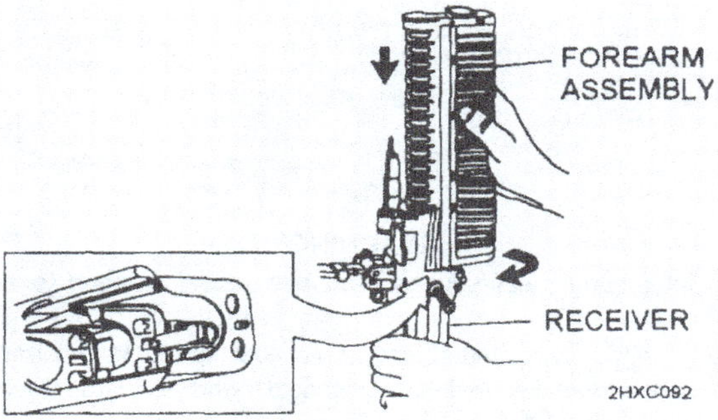

M60 TRIGGER MECHANISM GRIP ASSEMBLY OR M60D SEAR AND SAFETY HOUSING ASSEMBLY

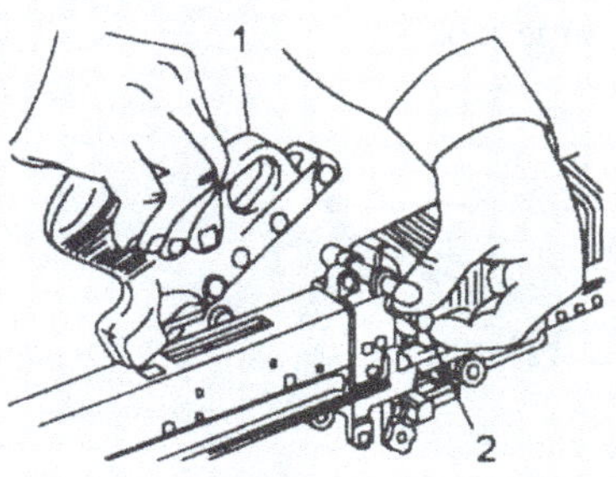

M60

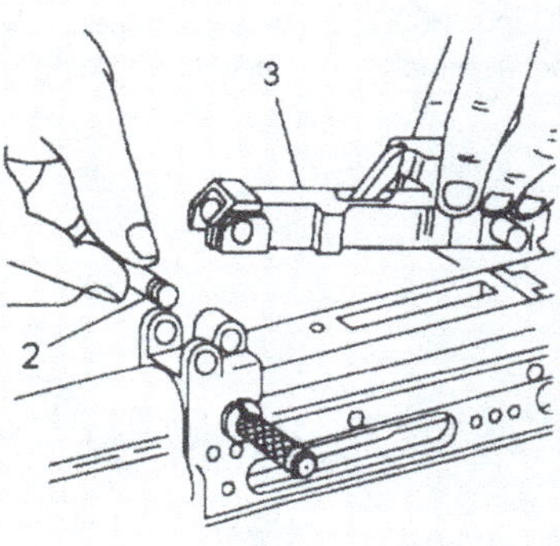

M60D

1 M60. Position trigger mechanism assembly (1) on bottom of receiver, align holes, and install front pin (2) from the left side.

M60D. Position sear and safety housing assembly (3) on bottom of receiver, align holes and install front pin (2) from the left side.

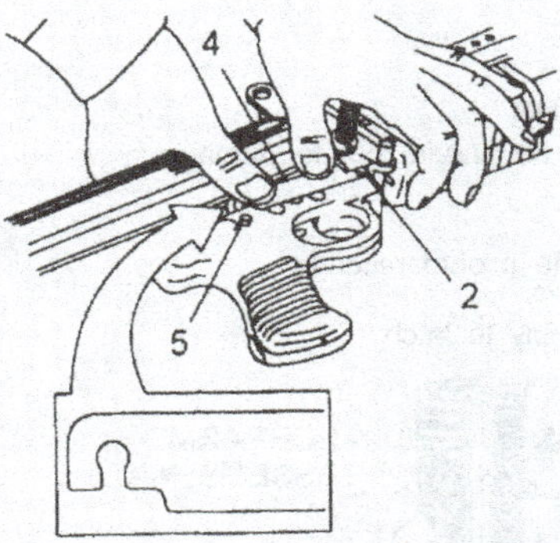

M60 2HXC095

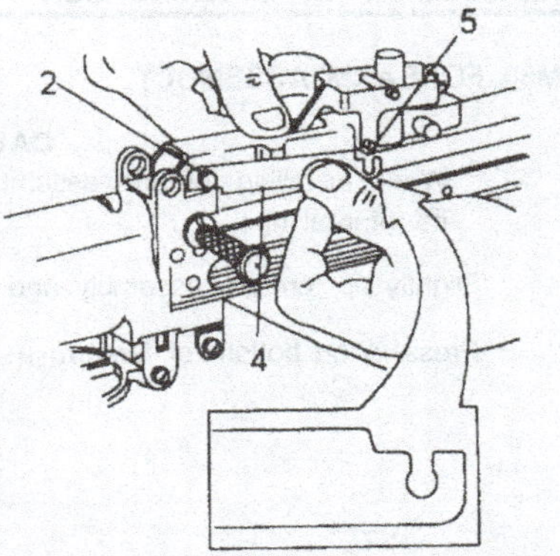

M60D 2HXC096

WARNING

Do not interchange leaf springs from one model weapon to another.

2 M60 and M60D. Slide slotted end of flat leaf spring (4) on front pin (2). The hooked end of flat leaf spring should bend outward. Push down and slide hooked end of flat leaf spring onto grooved pin (5).

BARREL AND BIPOD ASSEMBLY

M60 and M60D. Make sure barrel lock (3) is up (unlocked). Install barrel and bipod assembly (4). Push barrel lock (3) down to lock.

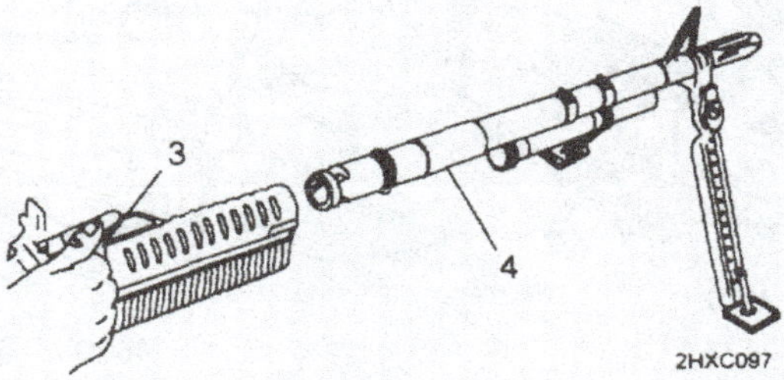

WARNING

To prevent possible bodily injury and aircraft damage, personnel should not stand the weapon on its barrel assembly when disassembling or assembling the weapon.

M60 D. Raise gun adapter (1), align holes, and insert quick release pin (2) to secure gun adapter to receiver.

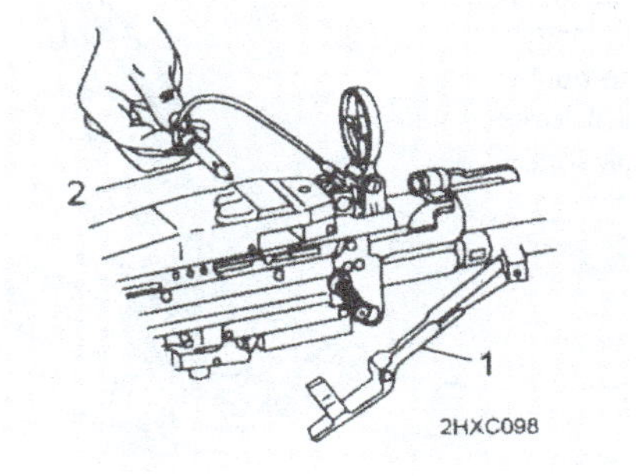

COVER ASSEMBLY AND M60 HANGER AND CARTRIDGE FEED TRAY ASSEMBLY OR M60D CARTRIDGE FEED TRAY ASSEMBLY

1 M60 or M60D. Position M60 hanger and cartridge feed tray assembly (1) or M60D cartridge tray assembly (2) on receiver.

2 Install torsion spring (3) and cover assembly (4). Make sure ends of torsion spring stick in holes of cover and receiver.

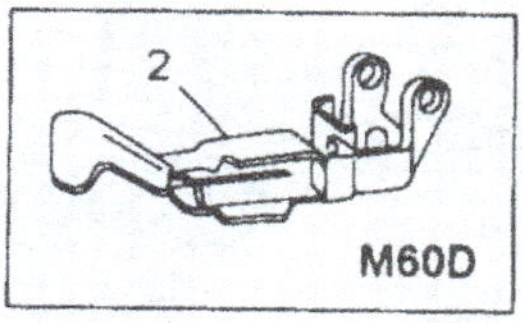

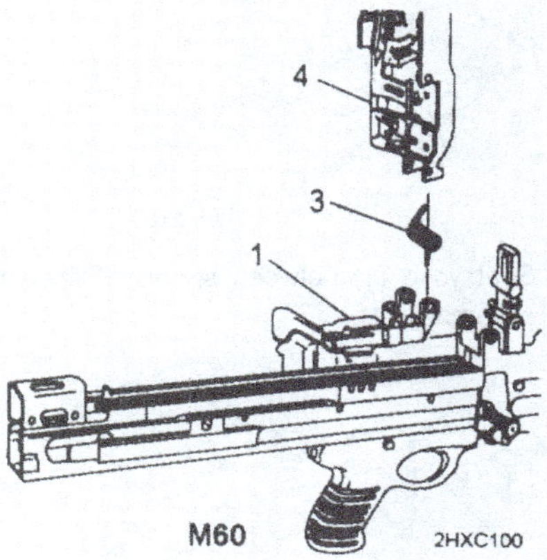

NOTE

Insert the hinge pin from one side and the hinge pin latch from the opposite side.

3 Apply slight pressure to line up torsion spring (3) and insert cover hinge pin (5). Install hinge pin latch (6) through cover hinge pin until it locks. Keep cover assembly open until machine gun is reassembled.

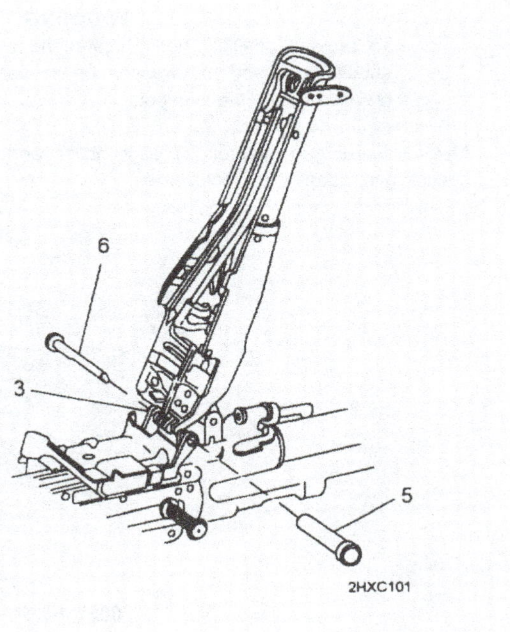

BOLT ASSEMBLY AND OPERATING ROD ASSEMBLY

1 Place yoke (1) of operating rod assembly (2) against spool of firing pin (3) in slot of bolt assembly (4) and push spool in direction of arrow.

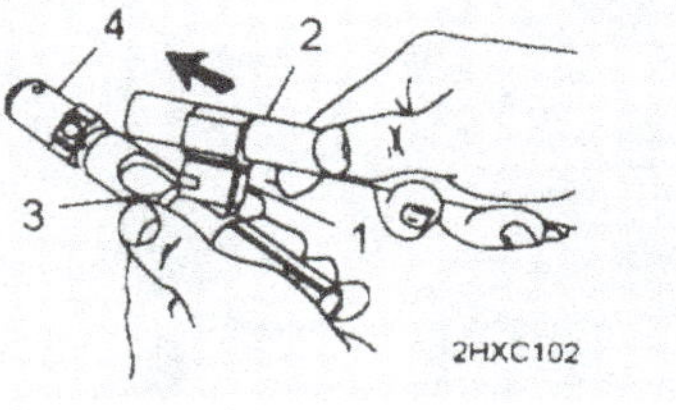

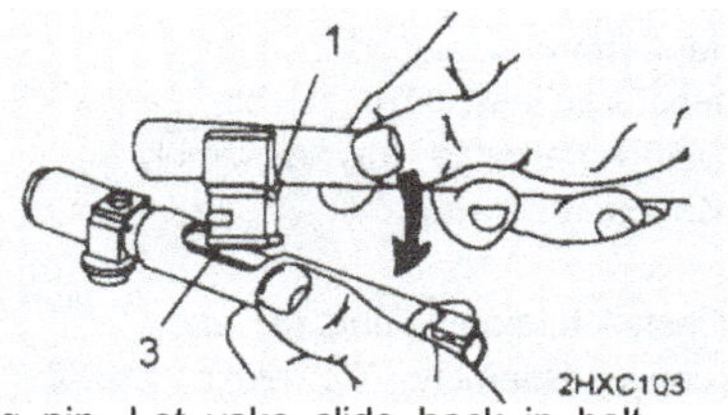

2 Seat yoke (1) between spools of firing pin. Let yoke slide back in bolt assembly (4).

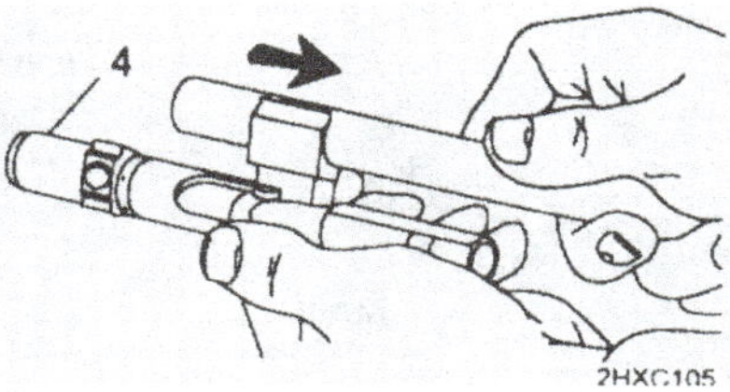

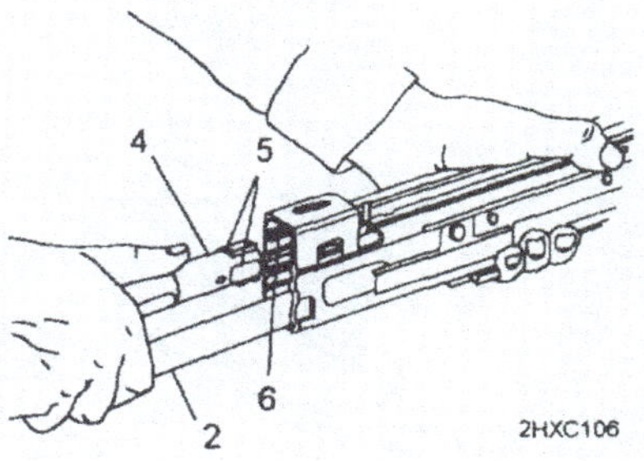

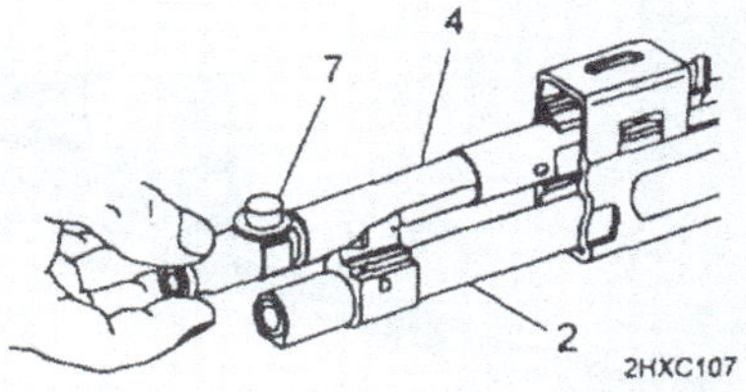

3 Start operating rod assembly (2) with bolt assembly (4) into weapon. Turn bolt assembly (4) so that stripping lugs (5) line up with upper rails (6). Push in until stripping lugs (5) engage rails (6).

4 Turn roller (7) straight up and push bolt assembly (4) and operating rod assembly (2) as a unit into weapon.

BOLT ASSEMBLY AND OPERATING ROD ASSEMBLY (cont)

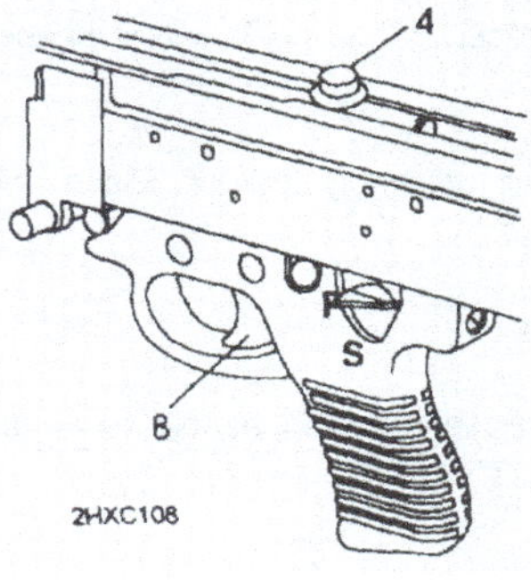

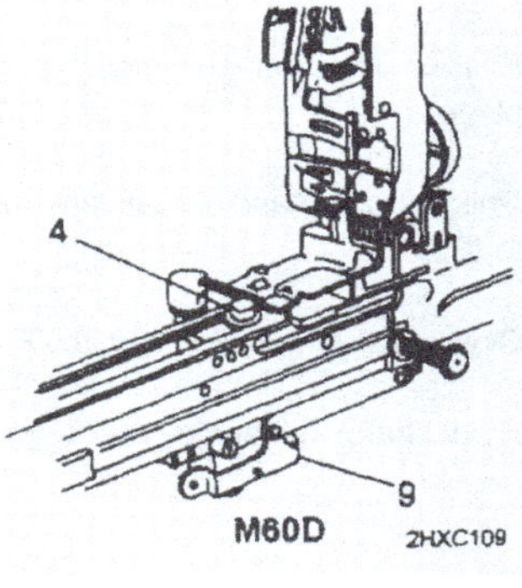

5 M60. Place safety to 'F'. Pull back on trigger (8) and hold while pushing bolt assembly (4) into receiver until it locks in place.

M60 D. Press in on 'F'. Pull down and hold sear assembly activator (9) while pushing bolt assembly (4) into receiver until it locks in place.

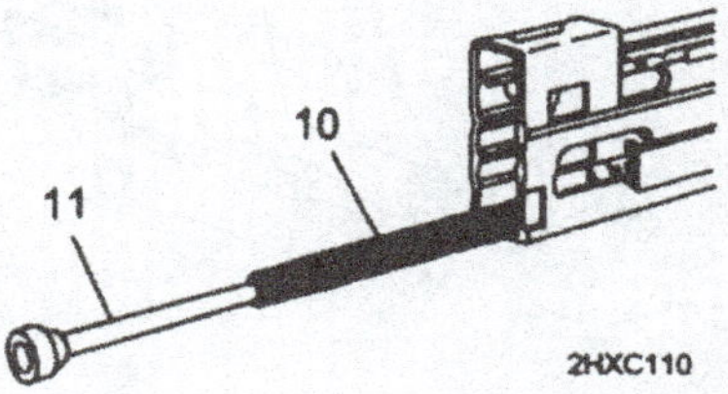

6 Install drive spring (10) and guide assembly (11) in rear of receiver.

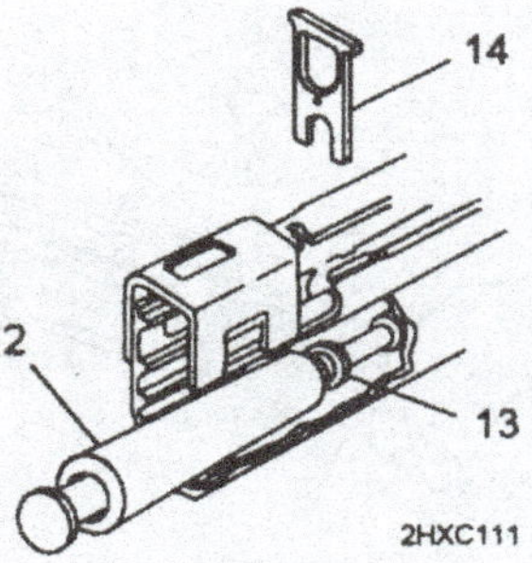

7 Insert buffer assembly (12) against end of guide assembly and push until groove (13) on buffer assembly lines up with slot for buffer retaining yoke (14).

6 Insert buffer retaining yoke(14).

M60 SHOULDER GUN STOCK

1 Position shoulder gun stock on rear of receiver and push until it snaps in place.

2 Pull cocking handle rearward to lock bolt assembly to rear. Place safety to ‘S’.

3 Close cover, move safety to ‘F’ position.

4 Hold charging handle, pull trigger, and ease charging handle forward.

M60 SPADE AND TRIGGER GRIP ASSEMBLY

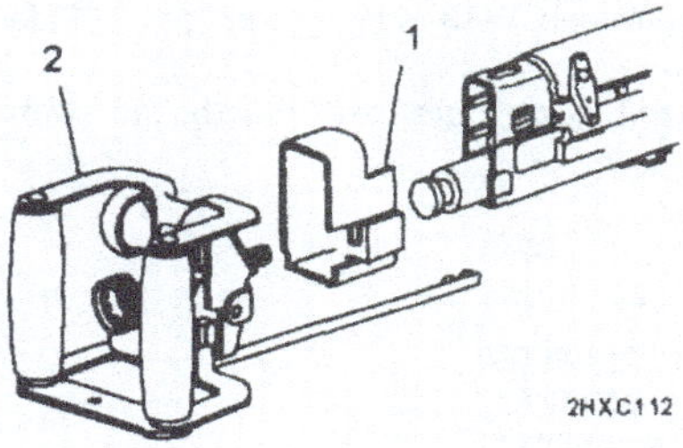

1 Install dust and moisture boot (1) and position grip and trigger assembly (2) on receiver.

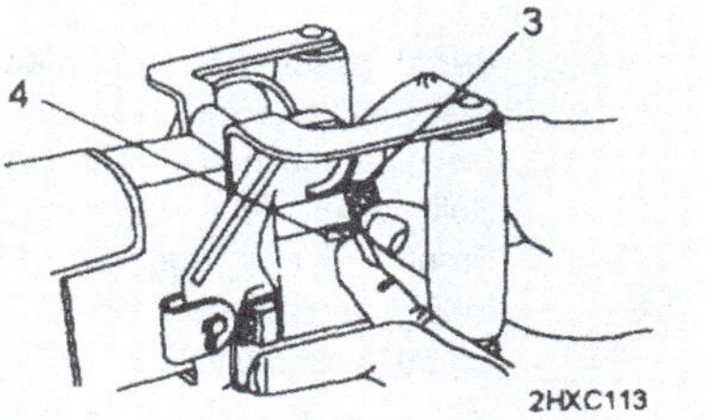

2 Turn knob and pin assembly (3) until clicking is heard. Then press and hold spring pin (4) and turn knob (3) until tight.

3 Release spring pin (4) and turn knob and pin assembly (3) until it locks.

M60 SPADE AND TRIGGER GRIP ASSEMBLY (cont)

4 Pull cocking handle rearward to lock bolt assembly to rear. Press in on ‘S’.

5 Close cover (5).

6 Snap end of sear assembly link and spring (6) on end of sear assembly activator (7).

7 Press safety in on ‘F’, hold cocking handle, pull trigger, and ease **bolt** forward.

END OF TASK

MAINTENANCE CHECKLIST 0028 00

Check the cleanliness and overall condition of the following:

- Barrel assemblies. Chamber bore and flash suppressor, gas cylinder, gas port, and gas piston.
- Receiver assembly. Operating rod guide, rails, and rear bridge rivets.
- Bolt assembly.
- Front and rear sights.
- M60 Shoulder gun stock.
- Bipod assembly.
- M60D Sear and safety housing assembly.
- M60 Trigger mechanism grip assembly.
- M60D Spade and trigger grip assembly.
- Cocking handle assembly.
- Cover assembly and M60 hanger and cartridge feed tray assembly.
- Cover assembly and M60D cartridge feed tray assembly.
- Forearm assembly.
- Blank firing attachment (BFA). Clean after firing 500 rounds or after each day's firing, whichever comes first. Remember BFA causes carbon fouling.
- Small arms sling.

FUNCTION CHECK 0029 00

Function check shall be performed after any maintenance or disassembly of the weapon.

Bolt assembly rearward? Good. Safety in 'S'? Good. Gun cleared? Good.

1 Load machine gun with dummy ammunition (WP 0005 00).

2 Pull trigger. Nothing should happen.

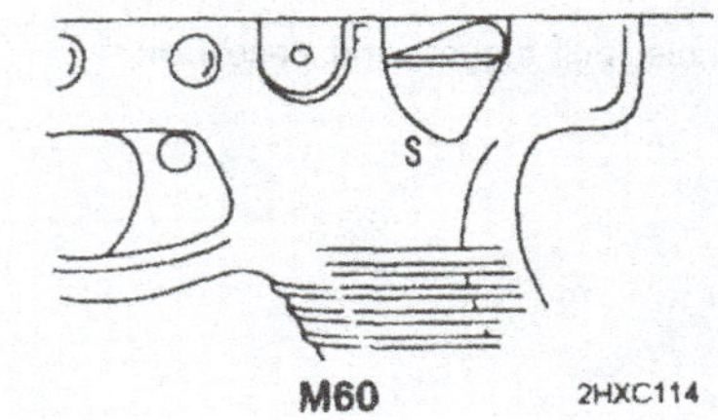

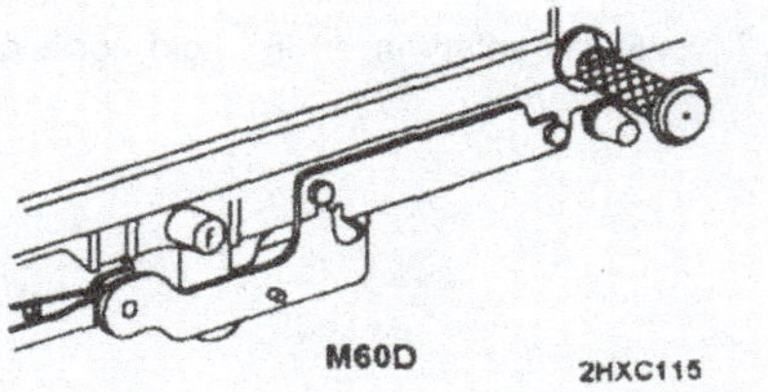

3 M60. Place safety to 'F'.
M60D. Press in on 'F'.

4 Pull trigger. Bolt assembly should move forward, strip and chamber a dummy round, and lock.

5 Pull cocking handle back. Dummy round should eject, and bolt assembly should lock to rear.

WARNING

After unloading dummy round, be sure the barrel is cleared.

6 M60. Place safety to 'S'.
M60D. Press in on 'S'.

7 Unload dummy ammunition and clear weapon (see WP 0009 00).

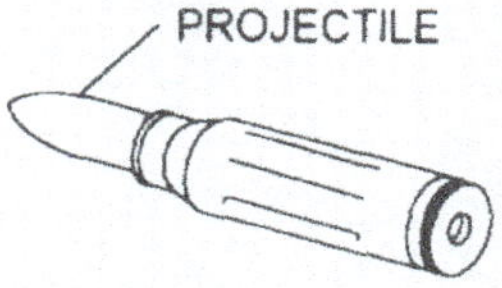

8 Be sure that projectile stays on dummy round.

9 Close **cover** and latch.

10 M60. Place safety to 'F'. Pull cocking handle, pull trigger, and ease bolt assembly forward.
M60D. Press in on 'F'. Pull cocking handle, pull trigger, and ease bolt assembly forward.

END OF TASK

CHAPTER 4

MAINTENANCE OF AUXILIARY EQUIPMENT

MACHINE GUN M122 TRIPOD MOUNT. Refer to TM 9-1005-245-13&P.

NIGHT VISION SIGHT AN/PVS-4. Refer to TM 11-5855-213-10 for maintenance.

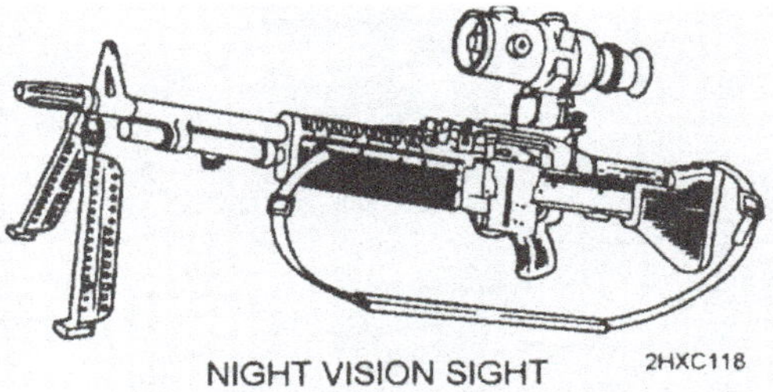

NIGHT VISION SIGHT

MOUNT PEDESTAL, MACHINE GUN, M6 AND MOUNT, MACHINE GUN M197. Refer to TM 9-1005-245-13&P

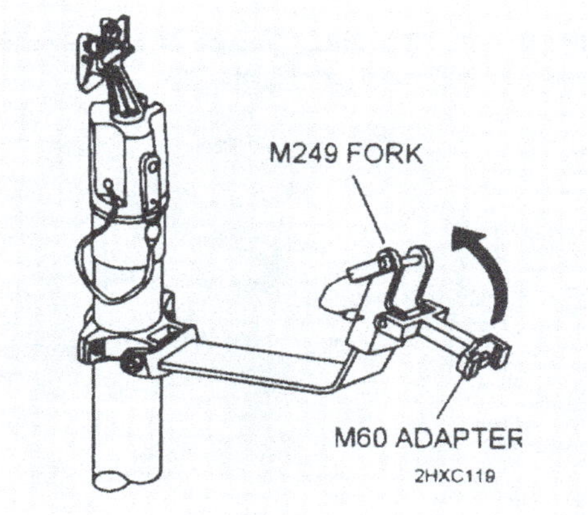

CHAPTER 5

AMMUNITION

WARNING

This is the only ammunition authorized for use in your machine gun. If it s not shown, it is not authorized.

AMMUNITION WHICH FAILS TO FIRE. Dispose of any ammunition, which fails to fire, according to authorized procedures.

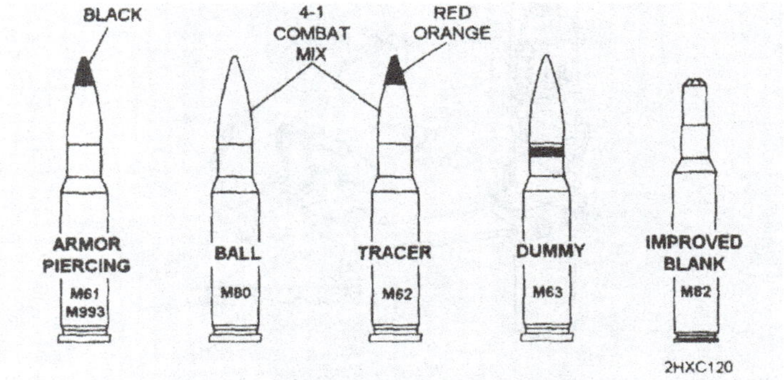

7.62-MM NATO AMMUNITION

CARE, HANDLING, AND PRESERVATION.

a. Do not open ammunition containers until the ammunition is to be used. Ammunition removed from the airtight containers, particularly in damp climates, is likely to corrode.

b. Protect ammunition from mud, dirt, and water. If the ammunition gets wet or dirty, wipe it off prior to use. Wipe off light corrosion as soon as it is discovered. Heavily corroded cartridges or cartridges which have dented cases or loose projectiles should not be fired.

c. Do not expose the ammunition to the direct rays of the sun. If the powder is hot, excessive pressure may be developed when the gun is fired.

d. Do not oil or grease ammunition. Dust and other abrasives collecting on oiled or greased ammunition will damage the operating parts of the gun, and oil on cartridges will produce excessive chamber pressure.

REFERENCES

REFERENCES 0032 00

SCOPE

This appendix lists all forms, field manuals, and technical manuals referenced in this manual.

CONSOLIDATED TABLE OF ALLOWANCES

CTA 8-100 .... Army Medical Department Expendable/ Durable Items

CTA 50-970 .... Expendable/Durable Items (Except: Medical, Class V, Repair Parts, and Heraldic Items)

FIELD MANUALS

FM 3-87 ....,...................... Nuclear, Biological and Chemical (NBC) Reconnaissance and Decontamination Operations (How to Fight)

FM 21-11 .... First Aid for Sholdiers

FM 21-40 .... NBC (Nuclear, Biological and Chemical) Defense

FM 23-67 .... Machine-gun 7.62-MM, M60

FIRING TABLE

FT 7.62-A-2	Machine Gun, 7.62-MM: M60 on Mount, Machine Gun: 7.62-MM, M122 and Machine Gun, 7.62-MM: M73 on Tank Combat, Full-Tracked: 105-MM Gun, M60 Series and Rifle, 7.62-MM: M14: Firing Cartridge, 7.62-MM, Ball, NATO, M59; Cartridge, 7.62-MM, Ball, NATO, M80; Cartridge, 7.62-MM: AP, NATO, M61/M993 and Cartridge, 7.62-MM; Tracer, NATO, M62

FORMS

DA Form 2028	Recommended Changes to Publications and Blank Forms
SF 368	Quality Deficiency Report

TECHNICAL MANUALS

TM 3-220	Chemical, Biological, and Radiological (CBR) Decontamination
TM 11-5855-213-10	Operator's Manual for Night Vision Sight, Individual Served Weapon, AN/PVS-4
TM 9-1005-245-13&P	Machine Gun Mounts and Combinations for Tactical/Armored Vehicles and Ground Mounting
TM 9-4933-273-12&P	Boresighting Equipment, Weapon Small Arms, M30

MISCELLANEOUS PUBLICATIONS

DA PAM 738-750	The Army Maintenance Management System (TAMMS)

COMPONENTS OF END ITEM

AND

BASIC ISSUE ITEMS LISTS

INTRODUCTION 0033 00

SCOPE

These work packages lists COEI and BII for the M60/M60D machine gun to help you inventory items required for safe and efficient operation of the equipment.

GENERAL

The Components of End Item and Basic Issue Items Lists are divided into the following lists:

Components of End Item (COEI). This list is for informational purposes only and is not authority to requisition replacements. These items are part of the M60/M60D machine gun. As part of the end item, these items must be with the end item whenever it is issued or transferred between property accounts. Items of COEI are removed and separately packaged for transportation or shipment only when necessary. Illustrations are furnished to help you find and identify the items.

GENERAL (cont)

Basic Issue Items (BII). These essential items required to place the M60/M60D machine gun in operation, operate it, and to do emergency repairs. Although shipped separately packaged, BII must be with the M60/M60D machine gun during operation and whenever it is transferred between property accounts. Listing these items is your authority to request/requisition them for replacement based on authorization of the end item by the TOE/MTOE. Illustrations are furnished to help you find and identify the items.

EXPLANATION OF COLUMNS IN THE COEI LIST AND BII LIST

Column (1) - Illus Number. Gives you the number of the item illustrated.

Column (2) - National Stock Number (NSN). Identifies the stock number of the item to be used for requisitioning purposes.

Column (3) - Description, CAGEC, and Part Number. Identifies the Federal name (in all capitol letters) followed by a minimum description when needed. The stowage location of COEI and BII is also included in the column. The last line below the description is the CAGEC (Commercial and Government Entity Code) (in parentheses) and the part number.

Column (4) - Usable On Code. When applicable, gives you a code if the item you need is not the same for different models of equipment.

Code	Used On
023	Model M60
G79	Model M60D

Column (5) - Unit of Measure (U/M). Indicates the physical measurement or count of the item as issued per the National Stock Number shown in column (2).

Column (6) - Qty Rqr. Indicates the quantity required.

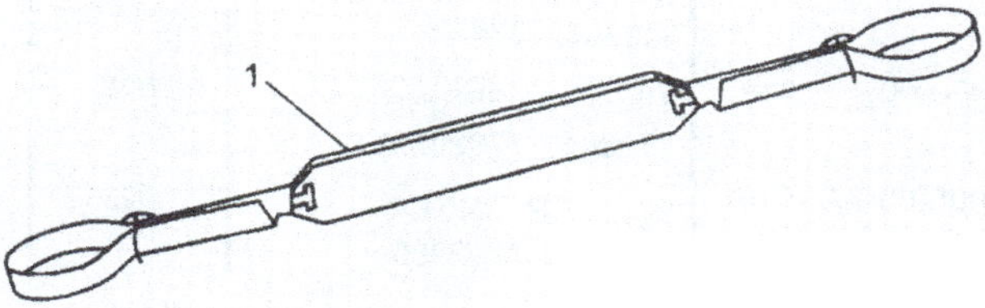

(1) ILLUS NO.	(2) NATIONAL STOCK NUMBER	(3) DESCRIPTION, CAGEC. AND PART NUMBER	(4) USABLE ON CODE	(5) U/M	(6) QTY RQR
1	1002-00-312-7177	SLING, SMALL ARMS (19204) 12003483	023	EA	1

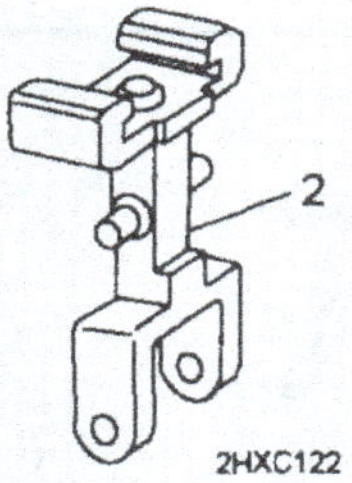

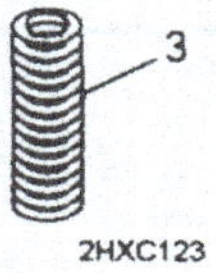

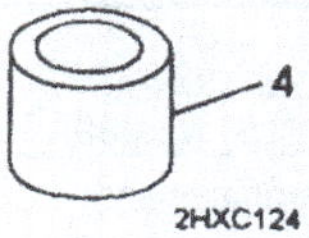

(1) ILLUS NO.	(2) NATIONAL STOCK NUMBER	(3) DESCRIPTION, CAGEC, AND PART NUMBER	(4) USABLE ON CODE	(5) U/M	(6) QTY RQR
2	1005-00-772-0194	ADAPTER ASSEMBLY (19204) 7792991	023	EA	1
3	5360-00-772-0193	HELICAL SPRING (19204) 7792987	023	EA	1
4	1005-00-772-0192	STUD, PLUNGER (19204) 7792986	023	EA	1

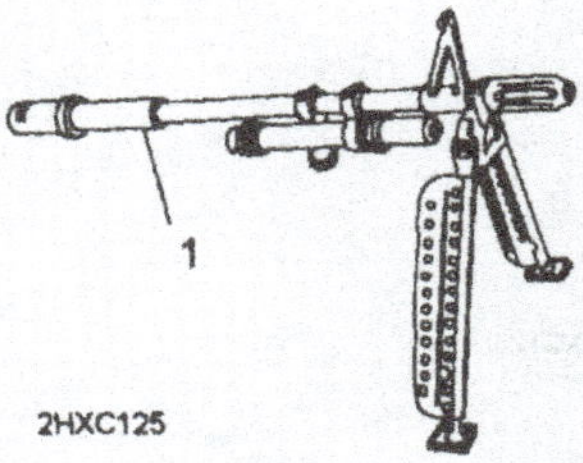

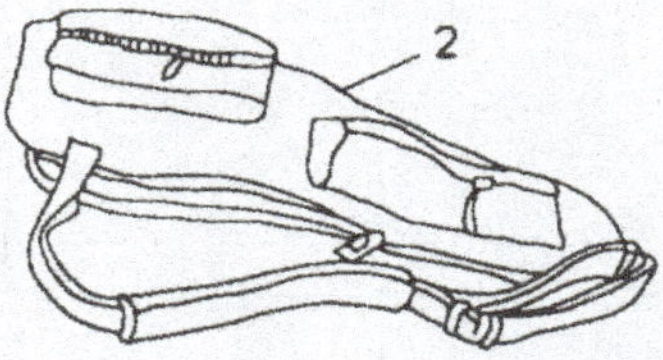

(1) ILLUS NO.	(2) NATIONAL STOCK NUMBER	(3) DESCRIPTION, CAGEC, AND PART NUMBER	(4) USABLE ON CODE	(5) U/M	(6) QTY RQR
1	1005-00-608-5001	BARREL AND BIPOD ASSEMBLY (19204) 7269027		EA	1
2	1005-00-791-5420	CASE, CARRYING, BARREL ASSEMBLY AND EQUIPMENT (19205) 7791009		EA	1

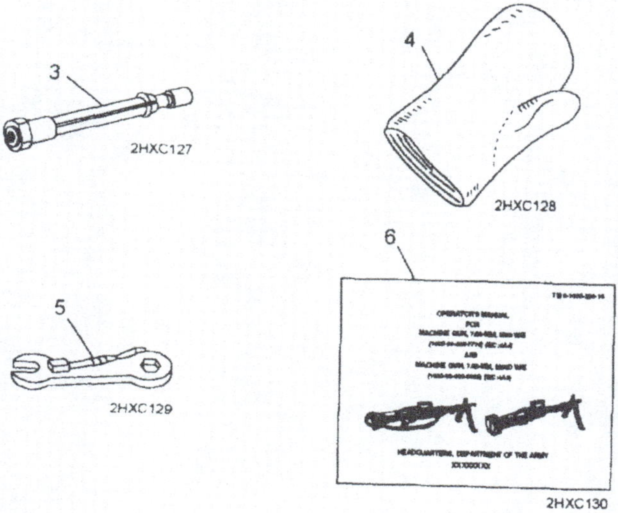

(1) ILLUS NO.	(2) NATIONAL STOCK NUMBER	(3) DESCRIPTION, CAGEC. AND PART NUMBER	(4) USABLE ON CODE	(5) U/M	(6) QTY RQR
3	4933-00-652-9950	EXTRACTOR, RUPTURED CARTRIDGE CASE (19205) 7790352		EA	1
4	8415-01-092-0039	MITTEN, HEAT PROTECTIVE (81349) MIL-M-11199		EA	1
5	5120-00-461-1075	WRENCH, SCREWDRIVER AND REAMER (COMBINATION) (19204) 8448458		EA	1
6		TM 9-1005-224-10		EA	1

ADDITIONAL AUTHORIZED LISTS

INTRODUCTION 0036 00

SCOPE

This work package lists additional items you are authorized for the support of the M60/M60D machine Gun.

GENERAL

This list identifies items that do not have to accompany the M60/M60D machine gun and that do not have to be turned in with it. These items are all authorized to you by CTA, MTOE, TDA, or JTA.

EXPLANATION OF COLUMNS IN AAL

Column (1) - National Stock Number (NSN). Identifies the stock number of the item to be used for requisitioning purposes.

Column (2) - Description, Commercial and Government Entity Code (CAGE), and Part Number (P/N). identifies the Federal item name (in all capital letters) followed by a minimum description when needed. The last line below the description is the CAGEC (in parentheses) and the part number.

EXPLANATION OF COLUMNS IN AAL (cont)

Column (3) - Usable On Code. When applicable, gives you a code if the item you need is not the same for different models of equipment.

Code	Used On
023	Model M60
G79	Model M60D

Column (4) - Unit of Measure (U/M). Indicates the physical measurement or count of the item as issued per the National Stock Number shown in column (1).

Column (5) - Qty Recm. Indicates the quantity recommended.

(1) NATIONAL STOCK NUMBER	(2) DESCRIPTION, CAGEC. AND PART NUMBER	(3) USABLE ON CODE	(4) U/M	(5) QTY RECM
	MTOE AUTHORIZED ITEMS			
5855-00-629-5334	SIGHT, NIGHT VISION AN/PVS-4 (60063) SMD850300-1	023	EA	1
1005-00-710-5599	M122 MACHINE GUN TRIPOD (19205) 7790723		EA	1
	CTA AUTHORIZED ITEMS			
1005-00-556-4174	BRUSH, CLEANING, SMALL ARMS: (BORE) (19204) 5564174		EA	2
1005-00-690-3115	BRUSH, CLEANING, SMALL ARMS: (CHAMBER) (19204) 7790452		EA	1
1005-00-350-4100	BRUSH, CLEANING, SMALL ARMS: (RECEIVER) (19204) 8848466		EA	1

(1) NATIONAL STOCK NUMBER	(2) DESCRIPTION, CAQEC, AND PART NUMBER	(3) USABLE ON CODE	(4) U/M	(5) QTY RECM
	CTA AUTHORIZED ITEMS (cont)			
1005-00-140-3515	FIRING ATTACHMENT, BLANK (19204) 12002981		EA	1
1005-00-793-6761	HANDLE ASSEMBLY: CLEANING ROD (19204) 7266115		EA	1
1005-00-726-6109	ROD, SECTIONS, CLEANING, SMALL ARMS (19204) 7266109		EA	5
	SWAB HOLDER SECTION, SMALL ARMS CLEANING ROD (19204) 7266110		EA	1

EXPENDABLE/DURABLE SUPPLIES

AND MATERIALS LIST

SCOPE

This work package lists expendable and durable items that you will need to operate and maintain the M60/M60D machine gun. This list is for information only and is not authority to requisition the listed items. These items are authorized to you by CTA 50-970, Expendable/Durable Items (Except Medical, Class V Repair Parts, and Heraldic Items), or CTA 8-100, Army Medical Department Expendable/Durable Items.

EXPLANATION OF COLUMNS IN THE EXPENDABLE/DURABLE ITEMS LIST

Column (1) - Item Number. This number is assigned to the entry in the list and is referenced in the narrative instructions to identify the item (e.g., Use wiping rag, (item 5, WP 0034 00).)

Column (2) - Level, This column identifies the lowest level ofmaintenance that requires the listed item.

C - Operator/Crew

EXPLANATION OF COLUMNS IN THE EXPENDABLE/DURABLE ITEMS LIST (cont)

Column (3) - National Stock Number (NSN). This is the NSN assigned to the item which you can use to requisition it.

Column (4) - Item Name, Descprition, Commercial and Government Entity Code (CAGEC), and Part Number (P/N). This column provides the other information you need to identify the item.

Column (5) - Unit of Measure (U/M). This code shows the physical measurement or count of an item, such as gallon, dozen, gross, etc.

(1) ITEM NUMBER	(2) LEVEL	(3) NATIONAL STOCK NUMBER	(4) DESCRIPTION, CAGEC, AND PART NUMBER	(5) U/M
1	C		Cleaner, Lubricant and Preservative: grade 2 (CLP)	
		9150-01-102-1473	1/2 oz bottle	EA
		9150-01-079-6124	4 oz bottle	EA
			(8 1349) MIL-L-63460	
2	C		Cleaning Compound, Solvent: Rifle Bore Cleaner (RBC) (81349) MIL-C- 372	
		6850-00-224-6656	2 oz can	EA
		6850-00-224-6657	8 oz can	EA
3	C		Lubricating Oil, Weapons (LAW) (81349) MIL-L-14107	
		9150-00-292-9689	1 qt can	EA

(1) ITEM NUMBER	(2) LEVEL	(3) NATIONAL STOCK NUMBER	(4) DESCRIPTION, CAGEC, AND PART NUMBER	(5) U/M
4	C		Lubricating Oil, Weapons: semi-fluid (LSA) (19204) 8436793	
		9150-00-889-3522	4 oz bottle	EA
5	C		Rag, Wiping: (cotton) (58536) A-A-531	
		7920-00-205-1711	50 lb bdl	LB
6	C		Swab, Small Arms Cleaning: (cotton) (19204) 5019316	
		1005-00-288-3565	(1000 per pkg)	PG

Subject | WP Sequence No.

A

B

C

Subject | WP Sequence No.

D

E

F

H

Subject **WP Sequence No.**

H (cot)

I

L

M

Subject **WP Sequence No.**

M (cont)

O

Subject **WP Sequence No.**

O (cont)

P

Subject **WP Sequence No.**

R

S

T

W

Made in the USA
Monee, IL
04 November 2026

67383097R00056